Displacement Measurement

Maria Teresa Restivo,
Fernando Gomes de Almeida,
Maria de Fátima Chouzal

Displacement Measurement

Book Series: Measurement of Physical and Chemical Quantities, Vol. 2

International Frequency Sensor Association Publishing

Maria Teresa Restivo,
Fernando Gomes de Almeida,
Maria de Fátima Chouzal
Displacement Measurement
Measurement of Physical and Chemical Quantities, Vol. 2

Illustrations: Leonor Zamith

ISBN-10: 84-617-3599-4
ISBN-13: 978-84-617-3599-0
BN-20141211-XX
BIC: TBMM

Contents

Foreword

Displacement Measurement constitutes one of the most challenging problems in Mechanical and Civil Engineering, crossing very different applications and scales, from biomechanical and mechanical components to long span bridges and aerospace structures.

The diversity of applications, test conditions, displacement ranges and accuracy demands have contributed to a continuous development in measurement technologies and sensors since the establishment of Experimental Mechanics to the present days. At the same time, the increasing potential of displacement measurement in the context of structural observation and health monitoring has fostered the dissemination of measurement techniques across different areas of knowledge.

Displacement Measurement is the second contribution from the authors for the book series Measurement of Physical and Chemical Quantities to be published by IFSA Publishing, focusing on techniques and sensors to measure physical and chemical quantities.

Addressed to researchers with different backgrounds, the book presents in an accessible and concise form the various physical principles and associated techniques for displacement measurement. These comprehend the most classical techniques, based on the measurement of variations with displacement of electrical resistance, capacitance and inductance, or else optical techniques using electromagnetic, acoustic or interference wave measurements, still with open fields of research and offering a wide range of applications. Complementarily and in a clear and systematic way, the authors discuss the advantages and limitations of the various techniques and provide information on the ranges of measurements and associated accuracy.

Written by researchers and academics from the University of Porto with extensive backgrounds in Mechanical and Electronic Engineering, the book *Displacement Measurement* is presented in a didactic, fluid and synthetic form and constitutes therefore an important instrument for the dissemination of knowledge in the various areas of research requiring the observation and health monitoring.

Elsa de Sá Caetano
Associate Professor at University of Porto

Preface

The present book entitled *Displacement Measurement* is the second title in the series 'Measurement of Physical and Chemical Quantities'.

This volume is devoted to the measurement of displacement considering its relevance in the engineering field. Invited authors also contribute to specific topics adding valuable perspectives.

The book starts by introducing the elementary concepts of position, displacement and distance. A brief overview of the history of length measurement is also included. Then the main traditional experimental methodologies for measuring displacement are described.

In the following chapters resistive and inductive transducers are introduced as well as digital encoders and transducers based in optical fibre. The chapters also include the basic instrumentation knowledge and techniques associated with the types of transducers described.

Chapter 1, *Elementary Concepts,* conveys the fundamentals of position, displacement and distance concepts and introduces a brief history of their units. A broad synthesis of the working principles used in the area is offered.

Chapter 2, *Potentiometer Transducers,* deals with the working principle of the most elementary transducer type for middle range displacements as well as its most relevant characteristics and highlights simple technical instrumentation details.

Chapter 3, *Inductive Transducers,* describes the fundamentals of the working principles of two inductive transducers, the LVDT and the Foucault Current types. It also details their fundamental specifications concerning signal instrumentation and conditioning as well as different techniques that should be applied in order to minimize the main sources of measurement error. Along the chapter the use of graphical representation is used to clarify many aspects of signal specification.

Chapter 4, *Digital Encoders,* explains how digital encoders of absolute and incremental types work. Their main characteristics are also mentioned.

Chapter 5, *Displacement Transducers Based in Optical Fibre Sensors,* is the contribution of the invited authors José Luís Santos, Luís Alberto Ferreira, Francisco Moita Araújo. The chapter offers insights on fibre optic based displacement sensors, providing a general classification of contact and non-contact types and describing sensor interrogation techniques and general characteristics.

The authors hope to have contributed to cover both the main concepts and the technical information on the topic of displacement measurement. This contribution has been clearly enriched by the involvement of the invited authors.

Chapter 1

Some Fundamental Concepts

1.1. Position, Displacement and Distance

The *position* of a rigid body is a *vector quantity* expressing the location of the origin of a Cartesian frame attached to the body $[i_1, j_1, k_1]$ in a base or reference Cartesian coordinate system $[i_0, j_0, k_0]$, with i_i, j_i and k_i representing the unit vectors of the coordinate systems main axes. Usually the rigid body centre of mass or other geometrically remarkable point is chosen as the origin of the frame $[i_1, j_1, k_1]$. So, the position of point O_1, origin of the Cartesian frame $[i_1, j_1, k_1]$, is described by a vector $p_1 = [x, y, z]^T$ when expressed in the base frame $[i_0, j_0, k_0]$.

Scalars x, y and z are the coordinates of point p_1 and represent *lengths* collinear with directions i_0, j_0 and k_0 and affected by a sign: positive if developed along the unit vector direction; negative if in the opposite direction. The length of vector p_1 is equal to its Euclidean norm and describes the distance from O_0, origin of the Cartesian base frame $[i_0, j_0, k_0]$, to point O_1. So, the length of a position vector or the *distance* between two points is a scalar quantity. This way of describing the quantities length and distance agrees with the definitions presented by the Oxford Dictionary:

Length is the measurement or extent of something from end to end;

Distance is the length of the space between two points.

The *displacement* of a rigid body is a *vector quantity* expressing a change in the body position. It may be obtained from the difference between the position vectors corresponding to the final and initial positions of a body motion. So, if at time t_i the body is at position $p_1(t_i)$ and at another time t_f, with $t_f > t_i$, the body occupies position $p_1(t_f)$, the body displacement d between time instants t_i and t_f is given by: $d = p_1(t_f) - p_1(t_i)$. The displacement only translates a position change giving no information on initial, final, or any intermediate position occupied by the body during its motion. In this way, motions with different origins and trajectories may induce the same displacement vector – equivalent displacements.

The length of the displacement vector, equal to the Euclidian norm of d, is a scalar quantity corresponding to the distance, in a Euclidian space, between the points $p_1(t_f)$ and $p_1(t_i)$. This is not necessarily the *distance travelled* by the body along its motion trajectory, which is given by *the length of the space travelled*.

So, a navigator that leaves a port and returns to it after a circumnavigation travel has a null displacement as points $p_1(t_i)$ and $p_1(t_f)$ are coincident, although he has travelled a distance of several thousand kilometers.

1.2. Angular Displacement and Orientation

When a rigid body is submitted to a motion such that all its points present the same displacement, it is said the body was the object of a *translation*. If different body points present different displacements then the motion is more complex. In general, when a rigid body moves its motion may be described by the superposition of two elementary movements: a translation and a *rotation*.

When a rigid body is submitted to a pure rotation – a motion with a null translation movement component – there is a set of points, not necessarily belonging to the body, presenting a zero displacement d. This set of points is located on a straight line – the *axis of rotation*. All body points not belonging to the axis of rotation perform circular trajectories – circular arcs – developed on planes perpendicular to the axis of rotation.

It is convenient to express the rotation movement by a quantity that, as in the pure translation case, is equal for all points of the rigid body. That quantity is the *angular displacement*. Its modulus corresponds to the plane angle swept by any moving body point (i.e., not belonging to the rotation axis) around the axis of rotation. So,

$$\theta = \frac{S}{r} \qquad (1.1)$$

S being the length of the circular arc trajectory performed by any moving body point, r is the distance between any point of that trajectory and the axis of rotation, and θ is the plane angle swept. The angle is said to be positive when corresponding to a movement developed anticlockwise, with the direction defined as positive with the axis of rotation pointing to the observer, and negative otherwise.

The *orientation* of a rigid body is equal to its angular displacement from a datum attitude. Common ways of expressing the orientation and the angular displacement are:

- Rotation matrix – matrix of the directional cosines of the body Cartesian frame $[i_1, j_1, k_1]$ relative to the base or reference Cartesian coordinate system $[i_0, j_0, k_0]$;

- Euler angles vector – vector composed by three angles corresponding to three non-commutative rotations, usually defined around axes i_1, j_1 and k_1, needed to take the body from the datum attitude, defined by $[i_0, j_0, k_0]$, to the final orientation;

- Rotation vector – vector obtained by the product of a unit vector with the direction of the axis of rotation by the plane angle θ.

1.3. Units of Length

Body displacements, as may be concluded from the previous sections, are expressed by one or more lengths (vectors p_1 and d) or by length ratios (angle θ, Euler angles) when referring to angular displacements. So, knowledge of the units used to express length is fundamental to the study of displacement measuring systems.

1.3.1. Traditional Units

Man has been searching for standards to express length since ancient times. The first units of length were based on human body dimensions or others related to it.

Nearly all cultures used the length of the human foot as a measuring unit. The *natural foot* (*pes naturalis* in Latin), an old unit based on the length of a bare foot, has a value of approximately 25 cm. This unit was replaced on the old Middle East civilizations by a longer foot, circa 30 cm, corresponding to a shod foot, as it was not convenient that someone had to take out his shoes in order to measure the length of a land property. It also had the advantage of being easily related to other natural units (1 foot = 3 *hands* = 12 *inches*). This unit was used by the Greeks and Romans. In England this unit was only changed after the Normand conquest, in 1066, giving place to the *modern foot* (1/3 of a *yard*, circa 30.5 cm). It is believed that it has been introduced by Henry I, King of England between 1100 and 1135. At the end of the XII century this unit was inscribed on the base of a column in St. Paul's Church in London so that everybody could see it, establishing one of the first official public standards (*de pedibus Sancti Pauli*). Since then this unit has not changed significantly. Henry I also redefined the yard as the length of a stick going from the point of his nose to the point of his right hand thumb with the arm in a horizontal position. The length of the oldest known *yardstick*, dating from 1445, is identical to the current definition with a difference smaller than 0.1 mm.

The USA, in 1959, and the United Kingdom, in 1963, redefined the inch, the foot and the yard relating these units with the *metre*, the unit of length of the *International System of Units*. As such, nowadays the official definitions of these units are:

- 1 inch, 1'' = 2.54 cm;

- 1 foot, 1' = 30.48 cm;

- 1 yard, 1 yd = 91.44 cm.

In other regions and contrarily to what happened in the United Kingdom and the countries that originated from the British Empire, this kind of temporal and territorial uniformity was not followed. In 789 Charlemagne ruled the uniformity of the measurement standards over all his Empire. Nevertheless, after his death in 814 there was a return to the old and diversified units. In France, for some centuries several kings unsuccessfully tried to impose a single standard in the kingdom. However, only the advent of the French Revolution enabled the start of that process during the reign of Louis XVI at the end of the XVIII century.

1.3.2. The Metre

The idea of defining a "natural" and universal standard for length measurement appeared in the XVII century. In 1670 the Abbot Mouton of Lyon, astronomer and mathematician, proposes as base unit the length of an arc with one arcminute of an Earth meridian (circa 1.852 m). Between 1670 and 1675 the French Picard, the Dutch Huygens and the Italian Burattini proposed for universal length unit the length of a pendulum with a half-period of one second (circa 0.994 m). However, as the Earth is not a perfect sphere this length changes with latitude. So, it would be necessary to define reference latitude.

In 1790 the French National Assembly adopts as standard the length of a pendulum with a half-period of one second at latitude of 45 degrees. In 1791 the French Academy of Sciences recommends to the French National Assembly the adoption of a fraction of the length of one fourth of an Earth meridian as the base for a new system. In the same year the French National Assembly accepts the meridional recommendation as the new base for a universal standard, unlike the previous one that had to be defined at given latitude. That was perceived at the time as a cause that could lead some countries to refuse the new standard. So, in the 19th of August 1791 Louis XVI ruled the beginning of a survey expedition to measure the length of the arc of meridian between Dunkirk and Barcelona, essential to the definition of the new standard. Given the turmoil during this revolutionary period the measurement of the arc of meridian was only concluded at the end of 1798.

In the 10th of December 1799 the French National Assembly decrees the metre as the new length standard unit defined as one ten-millionth of the arc of meridian between the North Pole and the Equator. A gauge was produced; a flat ruler of rectangular section built in platinum having a length between its extremities equal to the metre. This gauge, known since as the "Mètre des Archives", was deposited in the French National Archives.

After several efforts to internationalize the new standard an International Metre Commission takes place in Paris in August 1870 at the invitation of the French government appointing a preparatory committee. On May 20, 1875 seventeen of the twenty countries represented subscribed to the Metre Convention. This convention creates the three bodies that, to the present day, ensure the uniformity of standards of physical measures in the world:

- the International Bureau of Weights and Measures ("Bureau International des Poids et Mesures" – BIPM), permanent laboratory of metrology, installed at the Pavillon de Breteuil in Sèvres, a location near Paris;

– the International Committee for Weights and Measures ('Comité International des Poids et Mesures' – CIPM), the committee which directly controls the BIPM and is responsible for the preparation of decisions and recommendations;

– finally, the General Conference on Weights and Measures ('Conférence Générale des Poids et Mesures' – CGPM), a higher body which meets in Paris, currently every four years, gathering representatives of the signatories of the Metre Convention.

The first CGPM, which took place in 1889, sanctioned the prototypes of the metre. These prototypes – one international and several national – copies of the "Mètre des Archives" are X-section rulers made in platinum with 10 % iridium, with a length of 102 cm, with line marks at one centimeter from each end, Figure 1.1. The various national standards could not differ from the international standard by more than 0.01 mm when measured at ice melting temperature, as measured by a hydrogen thermometer. The prototypes were numbered and distributed by the signatory countries. The metre was then defined as the distance between the two line marks, at ice melting temperature, of the international standard deposited under the custody of the BIPM.

Figure 1.1. International metre and international kilogram of 1889.

In this first international definition the metre comes to be defined by the length measured from the prototype and no longer by its relation to the length of the arc of the terrestrial meridian. This was due, among other factors, to the discovery that the performed measurement underestimated that length by 0.2 mm due to an incorrect compensation of the flattening of the poles caused by the Earth rotation.

The definition of 1889 has been refined in 1927, at the 7th CGPM, having the metre been redefined as the distance, at 0 °, between the axes of the two lines marked in the platinum-iridium bar kept by the BIPM, and declared as the international meter standard by the 1st CGPM, with the bar at normal atmospheric pressure and supported on two cylinders of at least 1 cm in diameter, placed symmetrically on the same horizontal plane and 571 mm apart.

In 1960, the 11th CGPM, recognizing that the international prototype does not define the metre with enough accuracy and that the dematerialization of the standard was desirable in order to make it indestructible and reproducible, redefined the metre as the length equal to 1 650 763.73 wavelengths in a vacuum of the radiation corresponding to the transition between the levels $2p_{10}$ and $5d_5$ of the krypton-86 atom.

Furthermore, this 11th CGPM created the International System of Units (SI). It also determined the use of multiples and submultiples of that system, as well as the rules for writing the symbols.

The definition from 1960 remained until 1983, when the 17th CGPM adopted the definition currently in force:

The metre is the length travelled by light in vacuum during a time interval equal to 1/299 792 458 second.

Note that this definition sets the value of the speed of light in vacuum at exactly 299 792 458 ms^{-1}.

This definition, based on the speed of light in vacuum – a universal constant – was motivated by the advent and evolution of the LASER (Light Amplification by Stimulated Emission of Radiation), which allows to measure lengths with an accuracy much higher than previously attainable. Table 1.1 shows the different definitions and the uncertainties associated with them.

Table 1.1. Uncertainties associated with different definitions of the metre.

Date	Definition of metre	Accuracy
1799	Based on a fraction of one fourth of the Earth meridian	± 0.06 mm
1889	International Prototype of the Metre	± 0.002 mm
1960	Wavelength of the radiation emitted by krypton-86	± 0.000 007 mm
1983	Speed of light	± 0.000 000 7 mm
Today	Speed of light with high accuracy He-Ne LASER	± 0.000 000 02 mm

1.4. Physical Principles Used for Displacement Measurement

Displacement is a physical quantity whose knowledge is essential to the control of production equipment as well as for the operation of the most diverse types of automated commonplace systems. Moreover, the measurement of various physical quantities such as force, pressure, acceleration, flow, velocity, level, and others, is often performed indirectly from the measurement of displacements.

There are countless processes that can be used for measuring a displacement from the simplest, such as tape measure, to the most complex, such as LASER interferometers which can provide an accuracy of less than a nanometre. The processes leading to a translation of the displacement to be measured into an electrical signal that is directly linked with it are, however, increasingly used as they allow an easy integration of computer equipment in the measurement chain.

Thus, a number of physical principles which underlie the electrical displacement transducers most commonly used are presented below in a non-exhaustive manner.

- Variation of the electrical resistance with displacement

  When a mobile contact is made to slide over an electrical resistor, the observed resistance between one end of the electrical resistor and the movable contact is proportional to the displacement undergone by this one.

This is the principle used by *potentiometric transducers* which have measurement ranges that can vary between approximately 10 and 500 mm.

- Variation of capacitance with displacement

i) A capacitor formed by two parallel conducting plates has a capacitance inversely proportional to the distance between them – a displacement moving the conducting plates apart or closer translates into a change in their mutual capacitance which is inversely proportional to that displacement;

ii) A capacitor formed by two parallel conducting plates has a capacitance directly proportional to the projected area of one of the plates on the other – a displacement of one plate, in its plane, relative to the other causes a proportional change in their mutual capacitance.

Capacitive displacement transducers usually present measurement ranges between 0.05 and 20 mm.

- Variation of inductance with displacement

i) The inductance of a coil is a function of the magnetic permeability of its environment – the displacement of a ferromagnetic core changes that permeability causing a variation of the observed inductance.

This principle is often implemented with a differential assembly – two coils aligned to share a common core that when moved increases the inductance of one coil, simultaneously decreasing the inductance of the other. The coils are typically integrated in a bridge circuit, its unbalance being proportional to the core displacement. These *inductive displacement transducers* usually present measurement ranges from 1 to 500 mm.

ii) The mutual inductance between two coils is a function of the reluctance of the magnetic circuit coupling them – a relative movement between the coils or the displacement of a ferromagnetic coupling element causes a change of mutual inductance.

In the first setup (relative displacement between coils) this principle is the basis of inductive *eddy current transducers*. In these, one of the coils is embodied by the necessarily conductive surface of the body that undergoes displacement. These contactless transducers usually present measurement ranges between 0.25 and 30 mm.

The Linear Variable Differential Transformer (LVDT) is based on the second form of this principle. They present measurement ranges which typically lie between 0.1 and 500 mm.

* Variation of a wave time of flight with displacement

    When a pulse of electromagnetic or acoustic energy is projected onto a target a portion of that energy is reflected by it and returns to the source. The time of flight (roundtrip) that mediates between the projection of the pulse and the arrival of its reflection is proportional to the distance between the source and the target. A displacement of the target translates into a proportional variation of the time of flight.

    This principle is the basis of several displacement measurement systems which differ primarily in the energy type used. Thus, starting with the ones using electromagnetic pulses there is the RADAR (RAdio Detection And Ranging), which uses radio waves, and LIDAR (LIght Detection And Ranging), which uses LASER light. Both are used for measuring large displacements and distances, typically from a few meters to several kilometers. An interesting example of the use of LIDAR was its application in measuring the distance between the Earth and the Moon. In 1970 the Apollo astronauts placed reflective surfaces on the surface of the Moon. A LIDAR type device allowed the measurement of the average distance of 384.4 million meters, with an error of less than 30 mm.

    Acoustic pulses are used by SONAR (SOund Navigation And Ranging) developed for the detection of submarines and nowadays also used in the detection of fishing resources and seabed surveying. The use of ultrasound has allowed the implementation of ultrasound systems commonly used as an aid in medical diagnosis applications and detection of defects in structural components. Displacement transducers using ultrasound – in the air, in liquids or in solids, such as the *Temposonic* type – have measurement ranges between approximately 10 mm and 10 m.

- Interference between two beams of monochromatic light

  When a beam of monochromatic light, usually produced by a LASER, is divided into two equal beams, if the length of the path travelled by each of them until they reach a common target along the same direction is different, there is an interference phenomenon between the beams observable on the target. The interference phenomenon is due to the phase difference between the light beams when they reach the target. This is caused by differences in length of the paths, since the two beams produced from the same LASER beam are initially in phase. If one of the paths has a fixed length and the other a variable one – obtained by using a movable mirror, fixed to the body that undergoes the displacement – a displacement of half the wavelength of the light used causes a phase change of the same value. Thus, for example, if in the initial position the beams interfere constructively, being in phase, leading to a high luminous intensity zone on the target, the final position will show a destructive interference, since the beams are in phase opposition, causing a dark area on the same target site. Thus, a displacement of the mirror causes a temporal variation of the incident light intensity on the target. If photo detectors are integrated in this target, coupled to a circuit for counting the pulses produced, it is possible to measure displacements of the mirror with a resolution of half the wavelength of the light used. In this way, using an *interferometer,* it is possible to measure displacements up to about 100 m with a typical resolution of 300 nm. Using light intensity analogue sensors, this resolution can be improved, as the interference produces a sinusoidal intensity variation along the half wavelength. Using interpolation techniques on the produced signal an accuracy of less than 1 nm can be achieved. This type of displacement transducer is used as a primary standard by leading international metrology laboratories.

Finally, and in addition to the transducers presented in the previous paragraphs, *digital encoders* shall also be mentioned. These constitute a family of inherently digital displacement transducers. They consist of a ruler for measuring linear displacements, or a disk in the case of angular displacements. Being digital their measuring range is quantified, usually in a uniform manner. A binary code word is inscribed in each quanta, on the surface of the ruler or the disc, identifying unambiguously the corresponding position. According to the type of encoder binary words can be read using inductive, capacitive or optical sensors, the latter being the most common. Thus, a reading head containing the detectors and integral with the body undergoing the displacement, will provide as its output the binary words corresponding to each position occupied along the route. Measuring ranges between 10 mm and 30 m with resolutions between 10 and 0.1 μm are typically possible.

Chapter 2

Potentiometric Transducer

The potentiometric transducer is one of the simplest contact transducers designed for measurement displacements in the range 10 to 500 mm. The simplicity of operation principle and utilization gives it a good performance/cost relation. However, it presents some wear problems which may result in a degradation of its characteristics and reduce its working life.

2.1. Principle of Operation and Assembly Aspects

A typical potentiometric transducer consists of a fixed resistance, R_P, with a voltage supply and a wiper or sliding contact connected to the moving object.

According to the geometric shape of the resistance, the wiper motion can be linear, Figure 2.1a), of rotating type, Figure 2.1b) or a combination of both, helical type, Figure 2.1c).

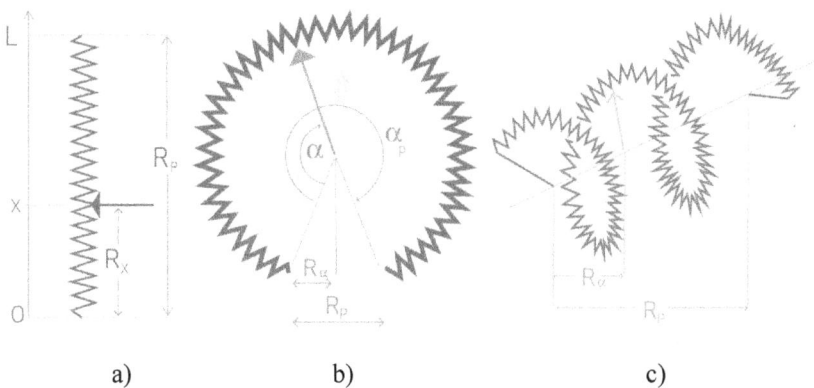

a)   b)   c)

Figure 2.1. Various forms of potentiometric transducers.

In all cases, the resistance value between the wiper contact and one of the fixed ends, R_x or R_α, is a function of the wiper position, and thus of the moving object whose displacement we want to measure, and the configuration of the resistance R_P.

The resistance RP consists of a wirewound coil or a thin conducting film, Figure 2.2. In the first case, the resistance wire is wound around an insulating core (glass, ceramic or plastic), and is itself also insulated except at the wiper contact points. In the second, the conducting film is made from a plastic coated with a carbon or metal conductive layer.

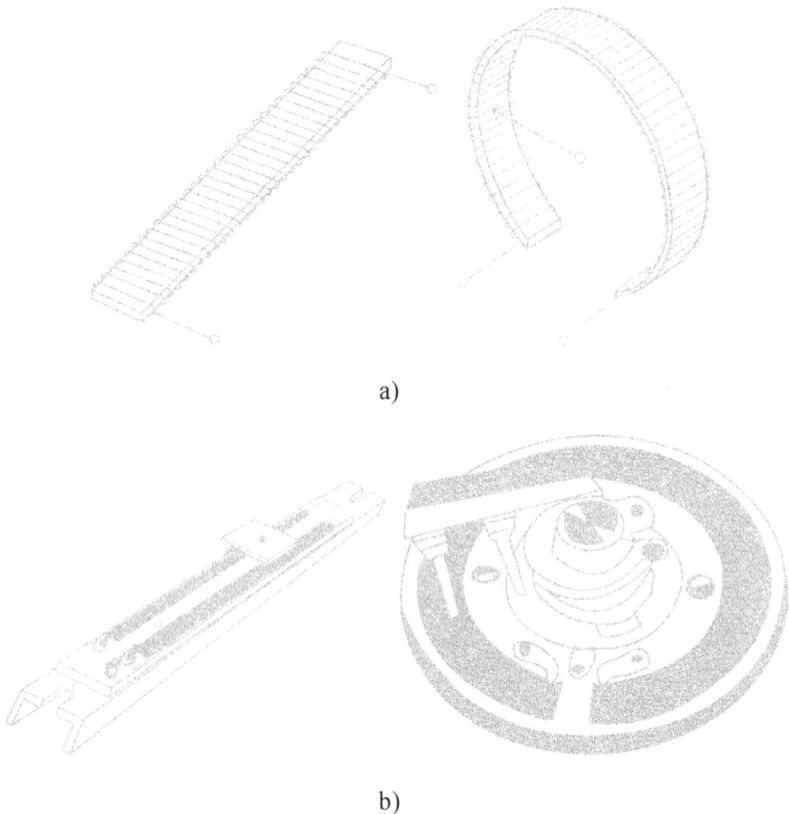

a)

b)

Figure 2.2. Potentiometric transducers: a) Wirewound; b) Conducting film.

The wound elements provide excellent stability with temperature and high capacity of power dissipation. The conducting film elements possess unlimited resolution, low friction and longer operating life.

24

There are also hybrid potentiometric transducers, in which the resistance assembly process combines the two previous technologies: one plastic track coated with a conducting layer is applied to a conventional coil element. These hybrid elements combine some of the best features of each technology, but with a higher cost.

Small non homogeneities in the structure or composition of the used materials or slight dimensional irregularities can lead to non-linearity in the relationship between the resistance value and the wiper position which will be reflected in the displacement measurement.

The wiper must ensure a good electrical contact without contact voltage drops and a low and stable contact resistance. An upper limit to wiper displacement speed is normally defined. The contact resistance depends on the pressure applied on the resistance by the wiper and on the contact surface state, being higher in the case of resistance conducting film. The wiper friction on the wirewound coil or conducting film causes wear resulting in operating life limitations.

2.2. Features of the Potentiometric Transducer

The potentiometer R_P is supplied by a voltage source, V_i, and the wiper contact creates a voltage divider, Figure 2.3.

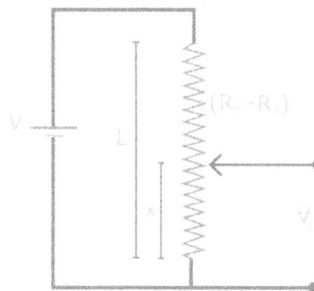

Figure 2.3. Potentiometric transducer.

The output voltage is given by:

$$V_o = \frac{R_x}{R_P} \cdot V_i \qquad (2.1)$$

Considering that the conducting element has a cross-sectional area A and a resistivity ρ, the resistance values R_P and R_x are given by the following expressions:

$$R_P = \rho \frac{L}{A} \qquad (2.2)$$

$$R_x = \rho \frac{x}{A}$$

Using (2.2) in (2.1), the relationship between the output voltage, V_o, and the displacement of the wiper, x, can be expressed as:

$$V_o = \frac{x}{L} \cdot V_i \qquad (2.3)$$

This linear relation is not achieved in practice due to the physical and geometric non uniformity of the resistance along its length L.

Furthermore, expression (2.1) requires infinite input impedance of the measuring device of the output voltage, V_o. If the measuring device presents a significant load to the potentiometric transducer, this is also responsible for deviations from the desired linear relationship. Figure 2.4 shows the new circuit considering the measuring device.

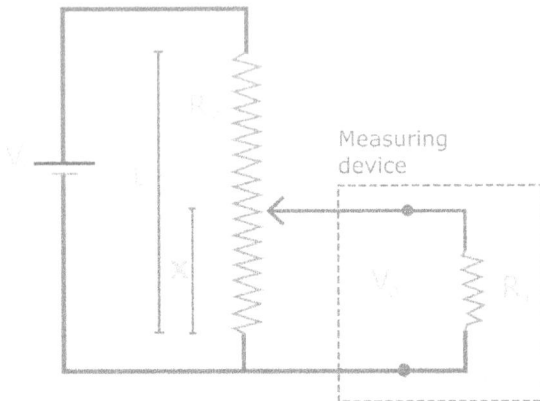

Figure 2.4. Loading effect on potentiometric transducer.

An analysis of the circuit leads to the following expression:

$$\frac{V_0}{V_i} = \frac{1}{\dfrac{L}{x} + \left(1 - \dfrac{x}{L}\right)\dfrac{R_P}{R_L}} \tag{2.4}$$

which indicates that the output signal of the transducer, V_o, is a nonlinear function of displacement x.

The deviations from linearity of the normalized transfer function of a potentiometric transducer in relation to the theoretical behavior ($R_L \to \infty$) for different values of R_P/R_L are shown in Figure 2.5.

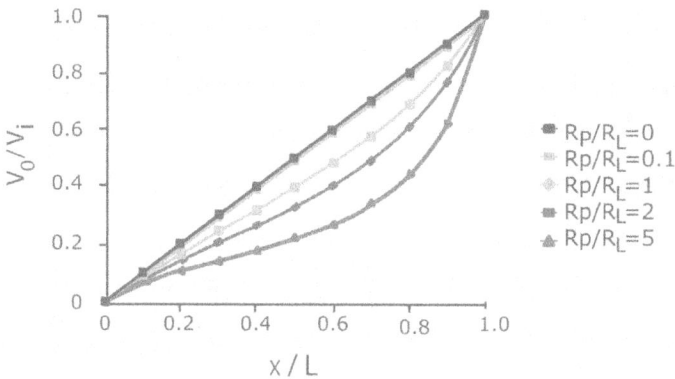

Figure 2.5. Influence of the input resistance of the measuring device on linearity.

Thus, the lower the resistance ratio RP/RL the better the linearity; typically a resistance ratio of one to ten is considered as a maximum acceptable value.

The choice of a potentiometer having a sufficiently low resistance with respect to R_L, improves the linearity, but conflicts with the requirement, always desirable, of high sensitivity.

Although sensitivity is directly dependent on the supply voltage, this cannot be increased indefinitely. Indeed, potentiometers have well defined power dissipation rates related to their heat dissipation capacity. If the power dissipated by the potentiometer is limited to P (W), the maximum supply voltage is given by:

$$V_{i(máx)} = \sqrt{P \times R_P} \qquad (2.5)$$

Analyzing expression (2.5), low values of R_P lead to low values of V_i and hence to a low sensitivity.

Example 2.1.

Consider that a 5 kΩ potentiometric transducer with a maximum travel of 100 mm is used for displacement measurement. The potentiometer is capable of dissipating 0.5 W of power. Which is the value of the potentiometer supply voltage in order to maximize the measurement sensitivity?

We will have:

$$V_{i(máx)} = \sqrt{0.5 \times 5 \times 10^3} = 50 \text{ V}$$

Thus, the choice of the potentiometer resistance value should result from a compromise between the loading effect of the measuring device (linearity or linearity error) and the required sensitivity.

For a given ratio R_P / R_L, the circuit shown in Figure 2.6 allows to minimize the loading effect of the measuring device, obtaining a better linearity.

The result obtained with this circuit is shown in Figure 2.7.

Another technique for minimizing the load effect of the measuring device uses an intermediate stage of amplification (between the transducer and the measuring device), characterized by having a very high input impedance and a very low output impedance.

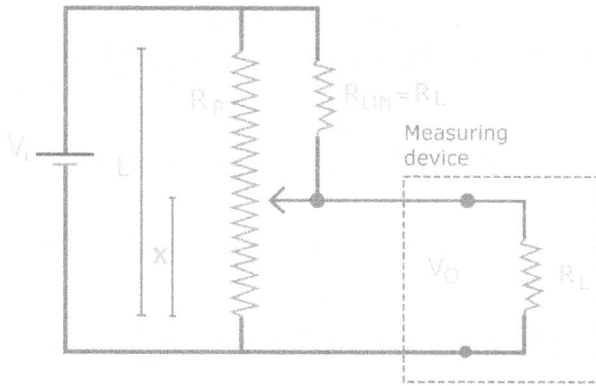

Figure 2.6. Circuit for minimizing the load effect of the measuring device.

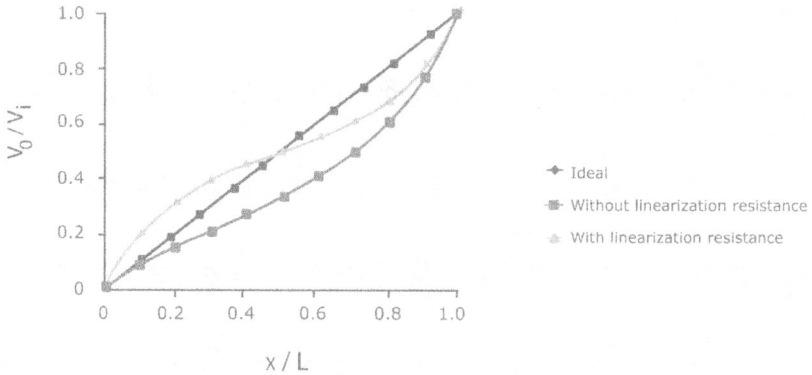

Figure 2.7. Influence of linearization resistance.

However, as the current measuring devices usually have high input impedance, the non-linearity associated with the loading effect is negligible.

2.3. Resolution

The resolution of a potentiometric transducer is strongly dependent on the assembly of the resistive element. Conducting film potentiometers present the highest resolution, theoretically unlimited, in practice only limited by the granular structure of the film. In wirewound potentiometers the resistance changes in a stepwise manner as the wiper moves from one turn to the adjacent turn along the helical resistance coil, which limits the resolution, Figure 2.8.

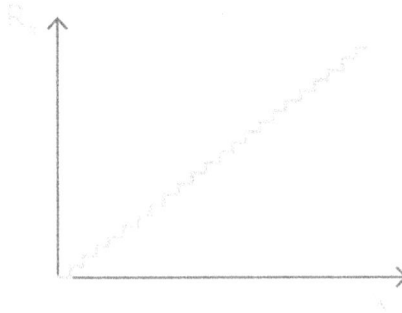

Figure 2.8. Resistance change in a wirewound potentiometric transducer.

Thus, in a potentiometric transducer in which the coil has 50 turns per mm, displacements of less than 20 μm will not be detected. This resolution corresponds to the worst value.

Indeed, when the wiper moves from one element to the adjacent element of the coil it passes through an intermediate position in which there is contact with both elements, as shown in Figure 2.9.

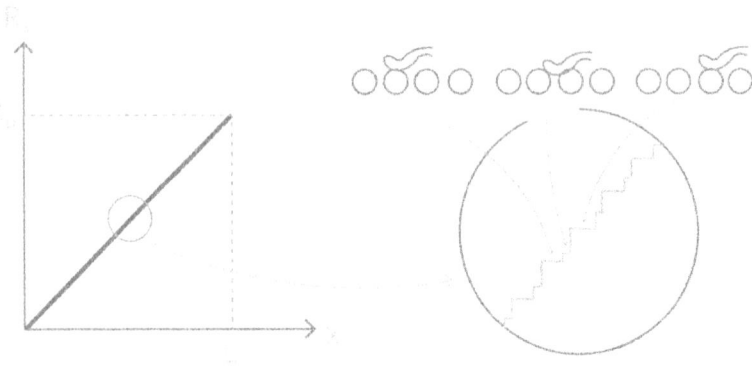

Figure 2.9. Wiper contact with one or two winding conductor elements.

In order to improve the resolution the number of turns per unit of length can be increased by reducing the diameter of the conductor coil wire.

However, this solution leads to an increase of the potentiometer resistance. When we select a wirewound potentiometric transducer within a given range, the choice of the resolution and the resistance values cannot be done independently. Furthermore, the use of conductor wire with a very small diameter leads to faster wear which may cause interruptions of continuity.

2.4. Synthesis of the Most Relevant Characteristics

The most relevant characteristics of potentiometric transducers are now briefly summarized.

Strong points:

- Ease of use
- Inexpensive
- Simplicity of operation

Weak points:

- The inertia and the friction present in the transducer mechanical parts preclude its use for dynamic measurements
- Wiper wear decreases the transducer reliability (transducer life)

Chapter 3

Inductive Transducers

3.1. Displacement Transducer of LVDT Type

3.1.1. Working Principle

A Linear Variable Differential Transformer – LVDT - is an electromechanical device that produces an electric output signal proportional to the displacement of its moving part – a ferromagnetic core. It comprises three coils, one primary coil and two secondary coils symmetrically wound relative to the primary one and connected together in a "series opposing" configuration. A cylindrical ferromagnetic core, of magnetically permeable material, free to move axially inside the windings and concentric with them, directs the magnetic flux through the secondary coils, Figure 3.1. The coil set is the static sensor element. The core is mechanically coupled with the moving target.

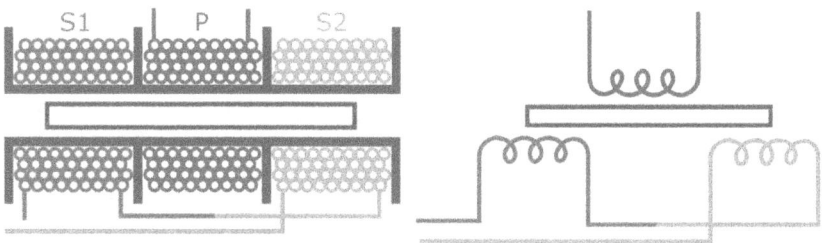

Figure 3.1. LVDT schematic layout.

The primary coil is excited by an alternating voltage (AC voltage), Figure 3.2, typically of sinusoidal type (typical values range from 3 to 15 Vrms @ 5 kHz). AC voltages are induced in the secondary winding (also of sinusoidal type) with frequency equal to that of the excitation and with amplitude varying with the position of the ferromagnetic core.

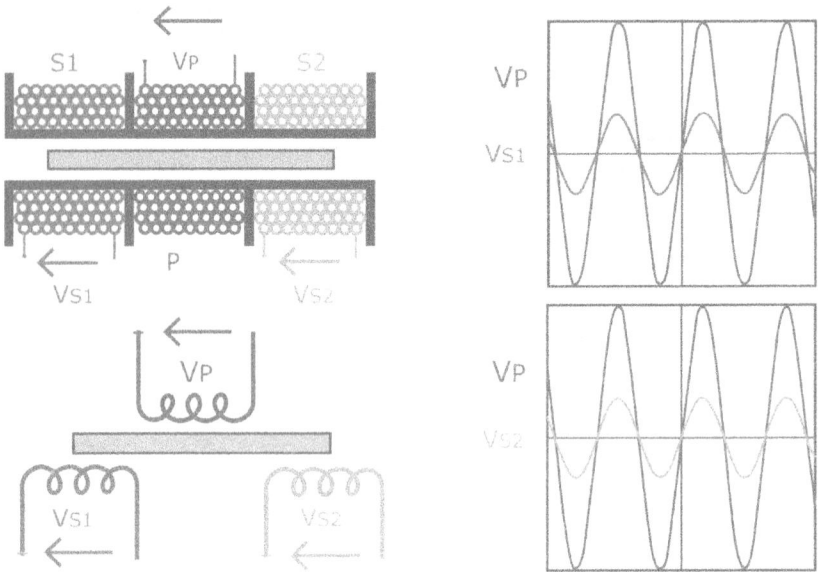

Figure 3.2. Primary excitation voltage V_P and induced voltages V_{S1} and V_{S2} in the secondary windings.

When the amplitudes of induced signals in the secondary windings are equal (theoretically, when the geometric center of the core coincides with the geometric centre of the secondary winding system) their difference is null (due to the type of electrical connections between them), Figure 3.3.

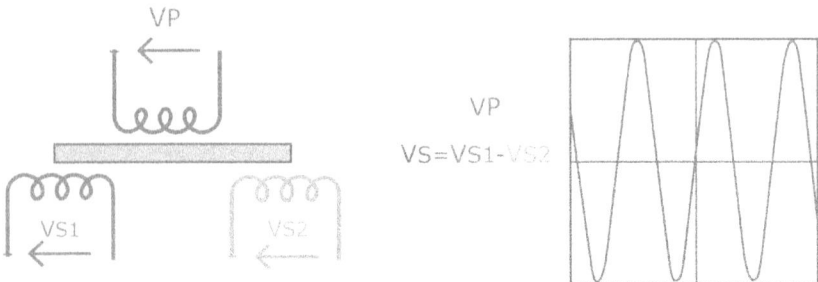

Figure 3.3. Core centre coincident with the geometric centre of the secondary winding system.

The core displacement to either side of the central position increases the mutual inductance between the primary and the secondary winding (whose coil is more affected by the core influence) and decreases the mutual inductance between the primary and the other secondary winding - whose coil is less influenced by the core position - which increases the signal in the former while decreasing the signal in the latter, Figure 3.4.

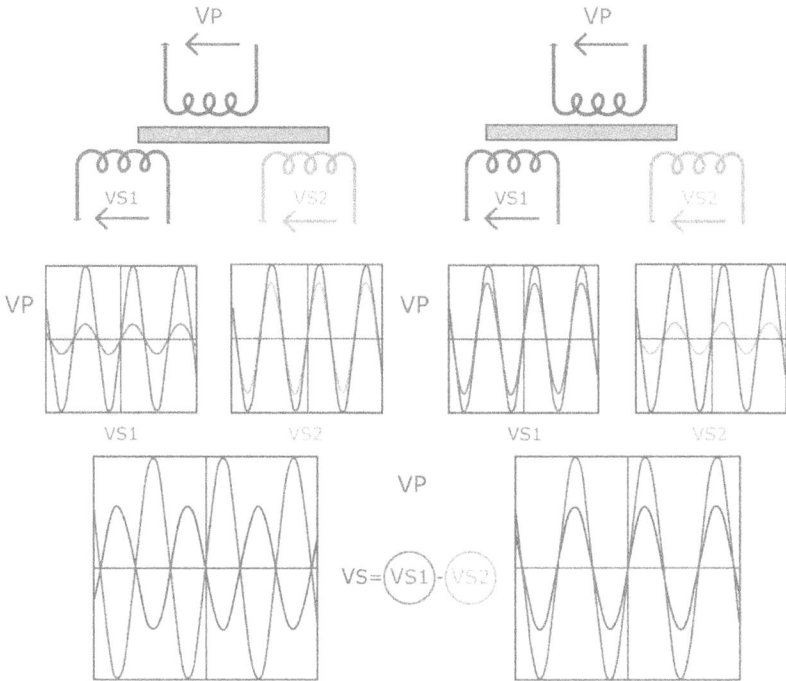

Figure 3.4. Secondary winding signals and their subtraction.

When the LVDT working principle was described, the fact was stressed that the two secondary windings are disposed symmetrically relative to the primary and connected together in a "series opposing" configuration, Figure 3.1. In fact, this simple mode of associating electrical coils allows the user to get the difference between two signals (equivalent to the sum of two sine wave signals in opposite phases), Figure 3.5.

The LVDT provides a differential output signal which is a function of the core position, i.e., it is a function of the core displacement. The position of the ferromagnetic core for which the output signal of the LVDT is zero is called "the LVDT null position". The differential output signal changes its phase by 180 ° when the core crosses the null position, from one side to the other, Figure 3.5.

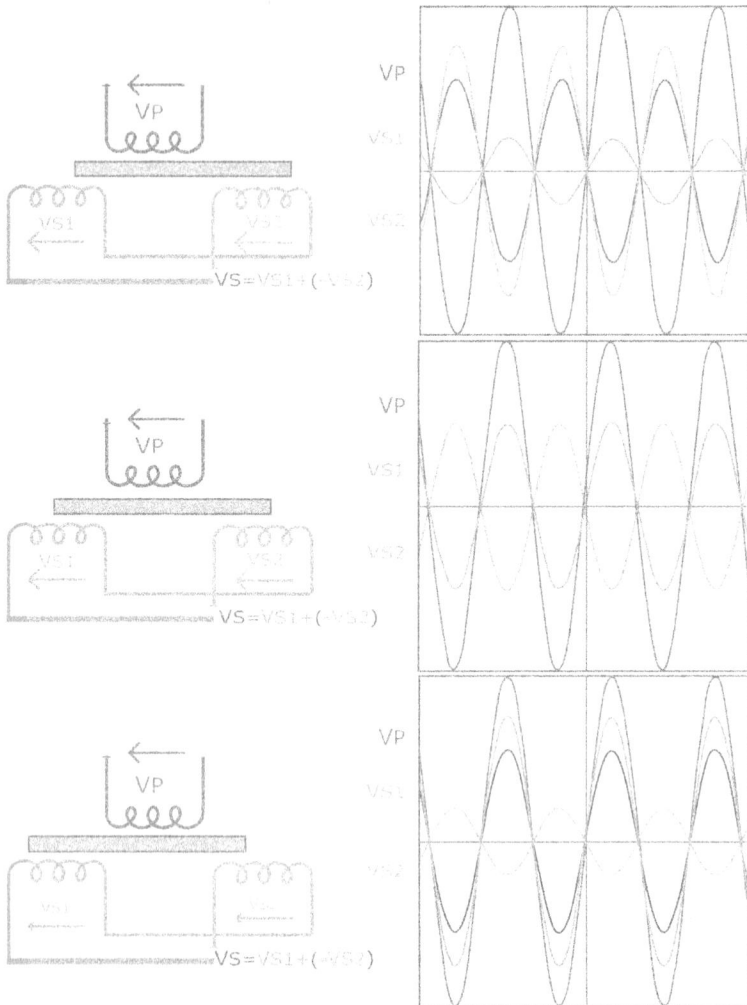

Figure 3.5. Secondary winding signals and their differential signal.

Usually "the LVDT null position" is very close to the geometric centre of its winding coil system. If they are not coincident the reason lies in very slight differences of the secondary coils geometry.

3.1.2. Working Principle and Electrical Circuit Equations

The electric circuit of an LVDT is illustrated in Figure 3.6. From the analysis of their primary and secondary circuits the following equations may be obtained relating for each circuit the electrical current, the voltage and the passive elements characteristics:

$$V_P = R_1 i_1 + j\omega L_1 i_1 + j\omega\left[M'(x) - M''(x)\right] i_2 \qquad (3.1)$$

$$\left[R'_2 + R''_2 + R_i + j\omega\left(L'_2 + L''_2\right)\right] i_2 + j\omega\left[M'(x) - M''(x)\right] i_1 = 0 \ (3.2)$$

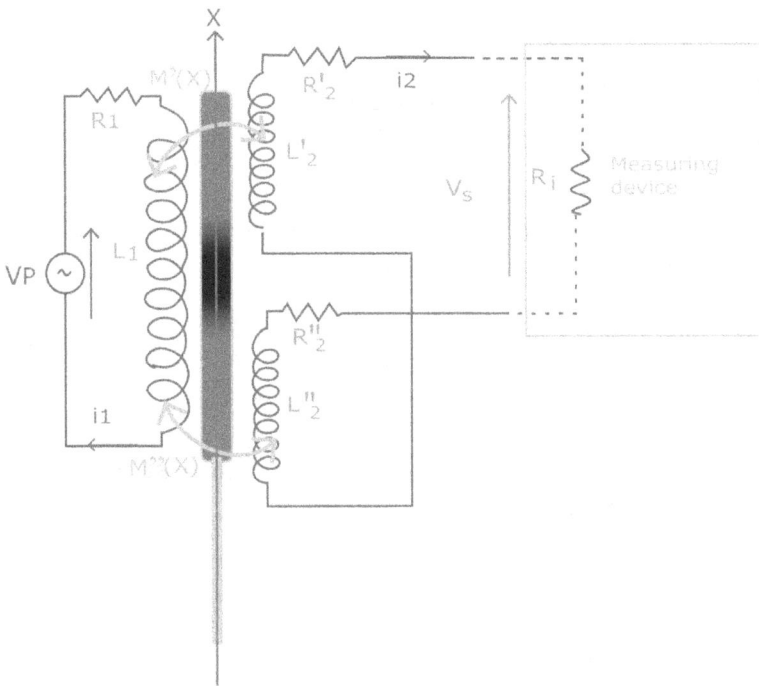

Figure 3.6. LVDT electric circuit.

If $R'_2 = R''_2 = R_2$ and $L'_2 = L''_2 = L_2$, then the voltage V_S across R_i is:

$$V_S = \frac{j\omega R_i \left(M''(x) - M'(x)\right)}{R_1(R_2 + R_i) + j\omega\left(L_2 R_1 + L_1(R_2 + R_i)\right) - \omega^2\left(L_1 L_2 + (M'(x) - M''(x))^2\right)} V_P$$

$$(3.3)$$

When the core is in "the LVDT null position", $M'(0) = M''(0)$ and $V_S \rightarrow 0$.

Considering the very high internal resistance R_i of the measurement equipment, equation (3.3) can be written as:

$$V_S = \frac{j\omega \left(M''(x) - M'(x)\right)}{R_1 + j\omega L_1} V_P \qquad (3.4)$$

that is, the voltage V_S is an independent function of R_i and becomes a linear function of the difference of the mutual inductance $(M''(x) - M'(x))$ coefficients. For high ω and low R_1 values, equation (3.4) can be written as:

$$V_S \approx \frac{M''(x) - M'(x)}{\omega L} V_P \qquad (3.5)$$

From equation (3.4) and considering the sensitivity definition and the type of functions $M'(x)$ and $M''(x)$, S is given by:

$$S = \frac{\Delta V_S}{\Delta x} \propto \frac{2\omega V_P}{\sqrt{R_1^2 + L_1^2 \omega^2}} \qquad (3.6)$$

For primary excitation of high frequency (as is the case of the typical values usually used) the sensitivity will be:

$$S \propto \frac{2V_P}{L_1} \qquad (3.7)$$

This leads us to conclude that the sensitivity becomes independent of the frequency of the primary excitation and is little dependent on temperature.

3.1.3. Conditioning Signal

In 3.1.1 it was stated and explained the working principle of an LVDT and it became clear that the output signal – the information about the ferromagnetic core displacement - has its frequency equal to the excitation frequency of the primary winding. At each moment this signal carries information on the position of the core according to its amplitude value. This is an example of an amplitude modulated signal.

The signal conditioning block will be analyzed next. It is important to get a simple and convenient final output signal for displacement monitoring and / or recording in the form of a DC signal.

Electrically, the LVDT provides an induced output signal V_S which translates the carrier wave signal (excitation signal) multiplied by the induced signal generated by the core displacement (modulating signal). So far only particular locations of the ferromagnetic core position were analyzed.

Now, it will be observed how a time varying displacement transforms itself into an output time varying signal V_S.

Figure 3.7 describes how a core displacement $x(t)$ with constant velocity, from one end of its measuring range to the other, is represented by the induced secondary output signal $V_S(t)$.

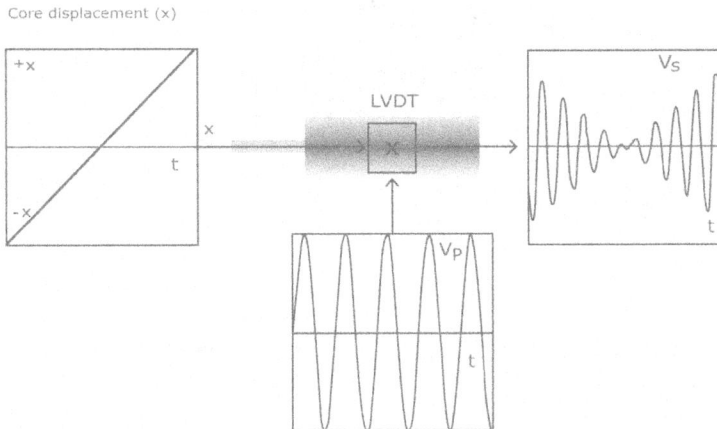

Figure 3.7. LVDT: a signal "multiplier".

The output signal V_S is amplitude modulated, i.e., information on displacement changes (modulating signal) is translated, instantaneously, by the carrier amplitude changes ($V_P(t)$), Figure 3.8. Whenever $x(t)$ crosses zero, the output modulated signal $V_S(t)$ changes its phase.

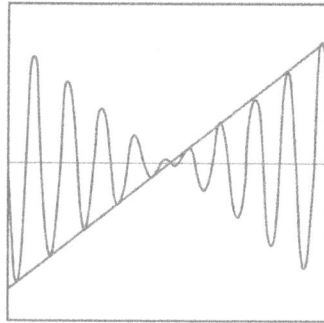

Figure 3.8. The modulated amplitude of the output signal carries information about the displacement.

Figure 3.9 presents another situation in which the output signal is different from the previous only in phase! In this case the core moves from one end of the measuring range to zero (the null position) and returns to the same end, i.e., it moves along half of the measuring range and back.

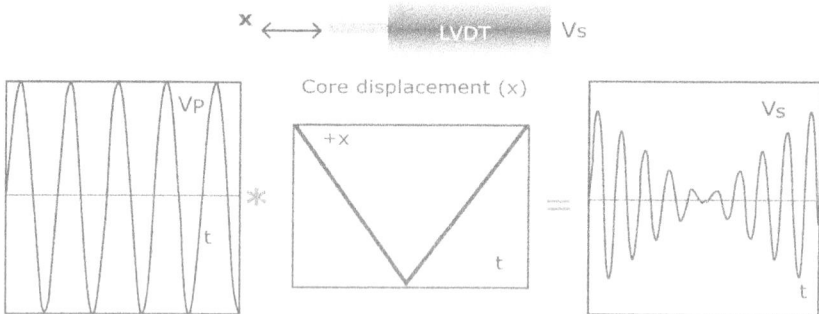

Figure 3.9. LVDT: the electromechanical signal "multiplier" for a new type of displacement.

And the displacement information is translated, instantaneously, by the carrier amplitude changes $V_S(t)$ presented in Figure 3.10.

If the output signals $x(t)$ of Figures 3.8 and 3.10 are varying in time very slowly the voltage $V_S(t)$ can be read on a voltmeter (AC) and its value can be calibrated in terms of displacement.

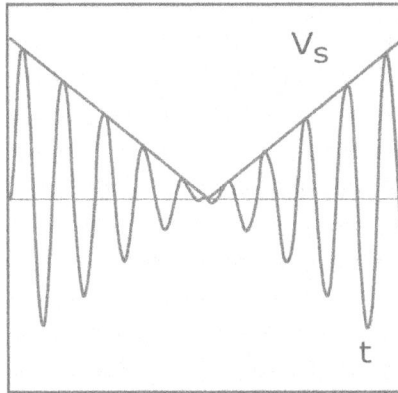

Figure 3.10. The amplitude modulation varies in accordance with the displacement.

However, it does not make it possible to detect the difference between the two distinct displacements represented in Figures 3.7 and 3.9. What is the difference between the output signals in these two cases? The output signals from Figures 3.8 and 3.10 differ in the signal phase when crossing the null position.

Figure 3.11 compares both situations by superposing the output signal on the excitation signal. In the first case, the excitation signal ($V_P(t)$) and the output signal ($V_S(t)$) differ in phase for symmetrical values of $x(t)$ relative to the transducer null position. In the second case signals ($V_S(t)$) and ($V_P(t)$) are always in phase.

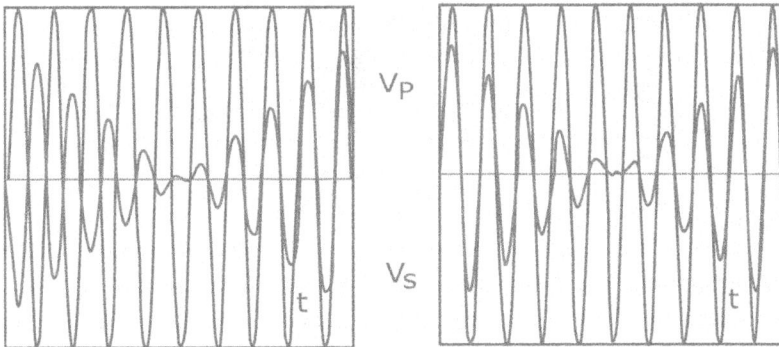

Figure 3.11. Output signals Vs(t) and carrier signal $V_P(t)$ compared in distinct displacements.

To get the complete information for the signals $x(t)$ represented in Figures 3.7 and 3.9, the output signals $V_S(t)$ in Figure 3.11 have to be converted into DC signals and also filtered.

If only a rectification is performed as shown in Figure 3.12, the information about the core position is not yet completely identified.

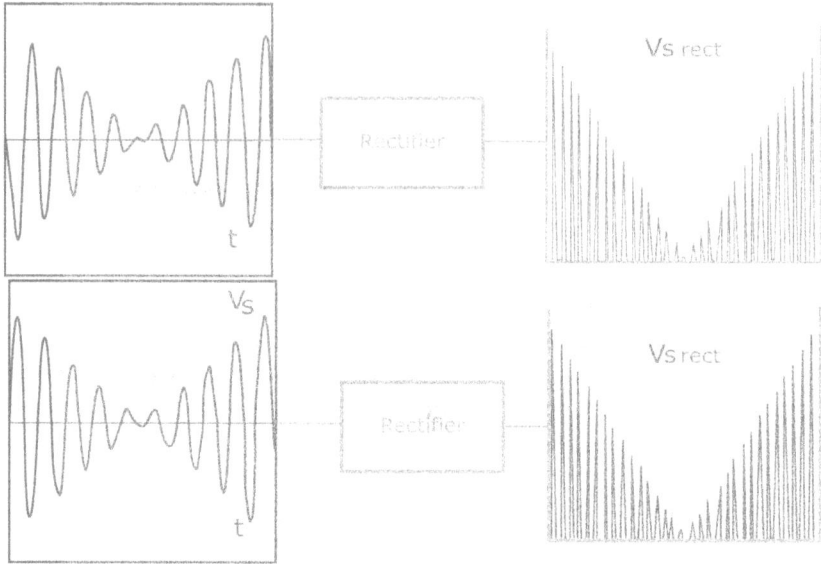

Figure 3.12. Rectified output signal (V_S) for the considered distinct displacements.

Phase detection is essential in the signal conditioning system of an LVDT. The amplitude modulated signal $V_S(t)$ must pass through a demodulator block with phase detection, enabling the conditioning system to define in which side of the null position the core is. Then, if the signal passes through a filter block (of low-pass type), the final signal reproduces the displacement signal $x(t)$ (relatively to the transducer null position), Figure 3.13.

Currently, the conditioning signal block offers all the necessary electronics to allow the user to get a DC voltage output by simply using a DC power supply. The market also offers displacement transmitters for LVDT offering current output signals of 4 ÷ 20 mA – named 2-wire displacement transmitters - especially used in industrial processes.

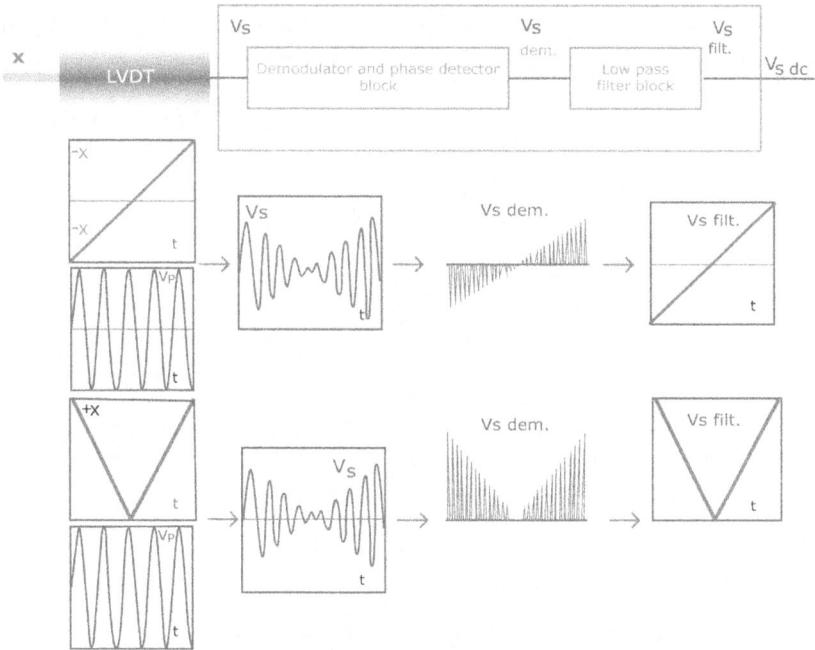

Figure 3.13. Input and output signals, demodulator block
with phase detection and filtering block.

3.1.4. Nominal and Measurement Range

The amplitude of the LVDT output signal is a function of the core position, offering a high linearity characteristic over a considerable range centered in the null position which is denominated measuring range. The signal demodulation with phase detection (phase-sensitive demodulator), also included in the conditioning module of the transducer, allows to know both the displacement value and the core position to the left or to the right of the transducer null position.

The relation between the output voltage and the displacement of the respective ferromagnetic core is named the LVDT characteristic. Figure 3.14.

When the sliding core ceases to influence one of the two secondary coils in either direction (it just influences the "other" secondary coil), saturation conditions have been reached and the measuring range (one of the operating characteristics specified by the manufacturer) is exceeded, as can be seen in Figure 3.15. The nominal range is wider and contains the measuring range.

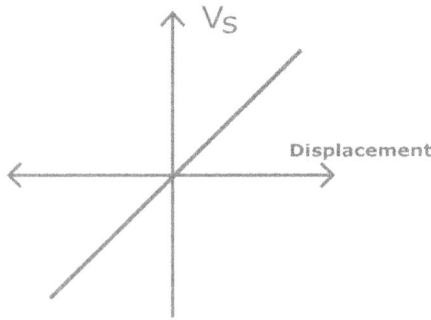

Figure 3.14. LVDT characteristic in the measuring range.

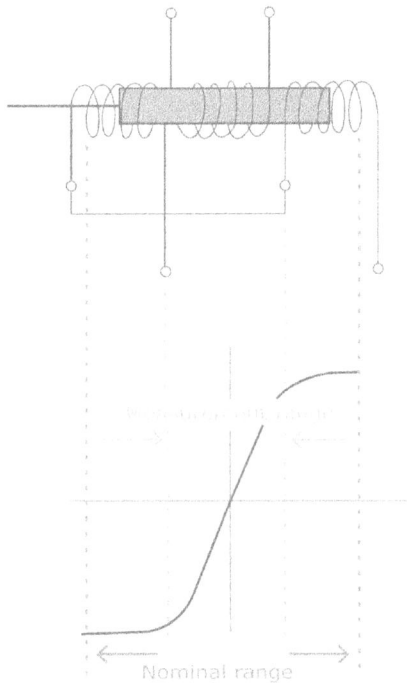

Figure 3.15. LVDT nominal and measurement (or measuring) range.

Note: the measurement range is frequently referred by manufacturers as nominal linear range.

3.1.5. Main Characteristics

Table 3.1 collects some of the general and universal characteristics of this type of transducer.

Table 3.1. LVDT main characteristics.

Frictionless operation	There is no contact between the sliding core and the sensing coils
Infinite resolution	The induction working principle is also associated with the electromagnetic coupling which makes the core displacement linearly related to its position. So, the transducer responds to any slight change of the position of its core. Therefore, the resolution is limited only by reading devices.
Mechanical wear of sensing elements	The LVDT has no parts that rub on each other, so it does not suffer from mechanical wear and has a long life cycle.
LVDT null position repeatability	The LVDT symmetric construction and its working principle lead to an inherent LVDT null position, giving to this transducer an excellent repeatability characteristic. Thus, an LVDT is an excellent null position detector for systems in closed-loop control.
Reduced sensitivity to transversal movements	An LVDT is not very sensitive to lateral movements when a radial displacement does not follow a straight path. The (undesirable) transverse sensitivity is generally less than 1 % of the longitudinal sensitivity.
Input /Output electrical isolation	Once the output signal (induced secondary signal) is electrically isolated from the input voltage (carrier signal), the signal conditioning does not require the use of isolation amplifiers.

Typical values for the primary supply are usually between 3 to 15 Vrms. Excitation signals typically have a frequency of 5 kHz and this value may extend up to 20 kHz. In the market there are LVDT transducers incorporating an oscillator block in its signal conditioning. In this case they only require an external DC power supply.

LVDT presents typical values of sensitivity between 10 to 500 mV/mm per Volt of supply voltage (usually sensitivity is presented as, e.g., 5 mV/V/0.001"). When the sensitivity of an LVDT is considered it is important its dynamic response (dynamic sensitivity) which is limited by the excitation frequency of the primary coil. This should be much higher than the frequency of the ferromagnetic core movement. The correlation between the excitation frequency of the primary and the core displacement frequency must be at least of the order of 10:1.

Measuring ranges exhibit a wide variety of values. At the lower limit typical values are ±0.1 mm and at the upper limit ± 250 mm. Different series available in the market cover values up to ± 25 mm at low ranges, and at high range values from ± 12 mm up to ±250 mm.

Linearity (or non-linearity or linearity error) is of the order of ± 0,5 % of measurement range. In lower ranges values can be ± 0.1 %. In higher measuring ranges linearity can be ± 1 %.

3.2. Eddy Current Displacement Transducer

3.2.1. Working Principle

An eddy current (or Foucault current) displacement transducer provides the distance measurement between the transducer head and a target of conducting material. Its sensor element usually integrates two coils. One, the active coil, is influenced by the increase or decrease of the distance to the conducting target and the other, the passive coil, aims to minimize the temperature variation effects in the transducer performance. These two coils are mounted in the transducer head and they are integrated in a measuring bridge, Figure 3.16, powered by an AC voltage supply of high frequency (≈1 MHz).

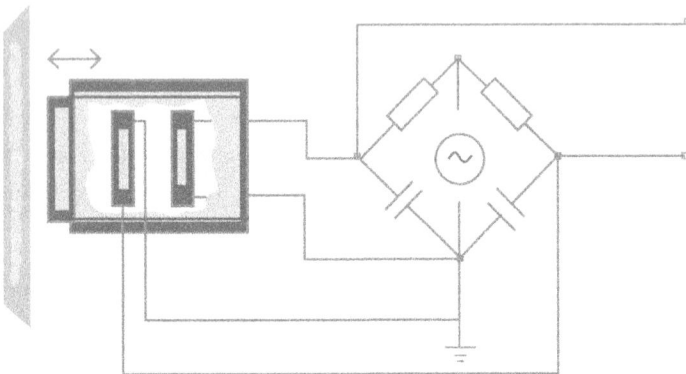

Figure 3.16. Eddy current displacement transducer.

The alternating current feeding the active coil creates on it a time varying magnetic field with a frequency equal to that of the power supply. The moving object (target) must, necessarily, have an electrically conducting surface (conducting target material or metallic surface). When the varying magnetic flux passes through the target eddy currents are induced on it, Figure 3.17.

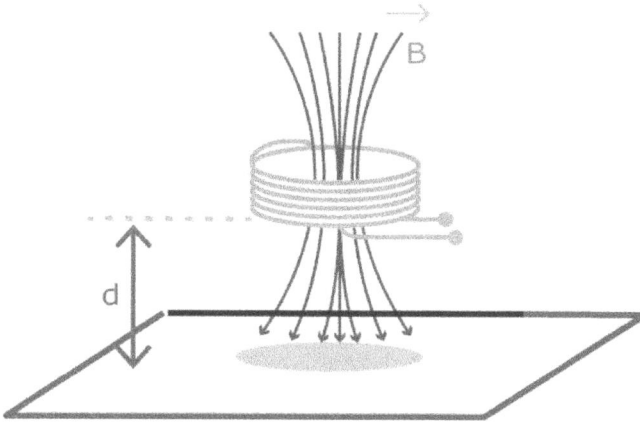

Figure 3.17. Induced eddy currents on the target surface.

These currents are mainly of surface type. Due to the typical operating frequency ($\approx$1 MHz) used in this type of transducers, induced eddy currents are strongly attenuated through the thickness of the conducting material. As the conducting surface of the moving target is approaching (or moving away) from the sensor head, the induced eddy currents become more (or less) intense; these currents will induce a reactive magnetic field resulting in a reduction (or increase) of magnetic flux through the coil and thereby changes on its active coil characteristics (inductance). And so, changes in the active coil inductance value cause a measurement bridge unbalance signal. This signal, after being demodulated and filtered, becomes a DC signal proportional to the distance between the target surface and the transducer probe.

3.2.2. Electrical Circuit Equations

The transducer can be electrically modelled by a weakly coupled air-core transformer: the coil constitutes the primary and the target is the (shorted) secondary, both coupled by an air-core. This means that an

RL series circuit represents its equivalent electrical circuit, Figure 3.18. Increasing the distance between the moving target and the active coil will increase its inductance and decrease the resistance and vice-versa as the target decreases its distance to the active coil. In fact, changes in distance between the inductances of the conducting material and of the active coil, produce a change in the coefficient of mutual induction (M) between these inductive elements, therefore causing an unbalance in the bridge equilibrium.

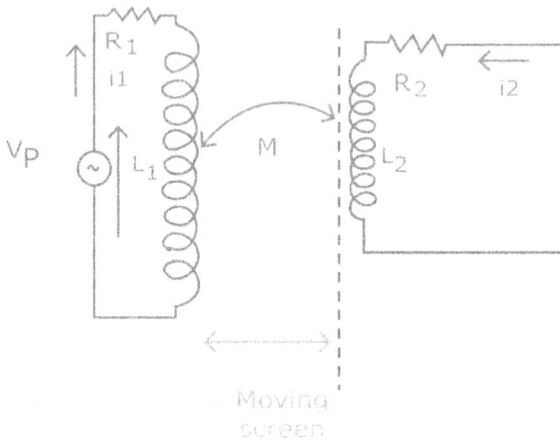

Figure 3.18. Equivalent electrical circuits (active coil and conducting target surface).

Equations for each of the electrical circuits are:

$$\left(R_1 + j\omega L_1\right) i_1 + j\omega M i_2 = V_{\text{P}} \tag{3.8}$$

$$\left(R_2 + j\omega L_2\right) i_2 + j\omega M i_1 = 0 \tag{3.9}$$

where V_{P} is the active coil excitation voltage which is integrated in the measuring bridge. The manipulation of those equations leads to:

$$\left(\left(R_1 + \frac{M^2\omega^2}{R_2^{\,2} + L_2^{\,2}\omega^2} R_2\right) + j\omega\left(L_1 - \frac{M^2\omega^2}{R_2^{\,2} + L_2^{\,2}\omega^2} L_2\right)\right) i_1 = V_{\text{P}} \tag{3.10}$$

which can be simplified to:

$$\left(R_{1eq} + j\omega L_{1eq}\right) i_1 = V_P, \qquad (3.11)$$

where:

$$R_{1eq} = R_1 + \frac{M^2\omega^2}{R_2^{\,2} + L_2^{\,2}\omega^2} R_2 \qquad (3.12)$$

and

$$L_{1eq} = L_1 - \frac{M^2\omega^2}{R_2^{\,2} + L_2^{\,2}\omega^2} L_2 , \qquad (3.13)$$

M is the coefficient of mutual induction, $M = k\sqrt{L_1 L_2}$, where k is the coil coupling coefficient. For a good conducting material $R_2 << \omega L_2$, whereby:

$$\frac{M^2\omega^2}{R_2^{\,2} + L_2^{\,2}\omega^2} \cong k^2 \frac{L_1}{L_2} \qquad (3.14)$$

And equations (3.12) and (3.13) become:

$$R_{1eq} = R_1 + k^2 \frac{L_1}{L_2} R_2 \qquad (3.15)$$

and

$$L_{1eq} = L_1 + \left(1 - k^2\right), \qquad (3.16)$$

where the coupling coefficients, k, between coils is a function of the distance between them.

Changes in the value of the equivalent inductances create unbalance in the measuring bridge. After demodulating and filtering the bridge output a DC signal gives the distance between the target and the active coil.

3.2.3. Main Characteristics

The eddy current transducers in the market have measuring ranges from 0.25 to 30 mm. In this type of transducer the range is related to the coil diameter of its active sensing element, with a ratio (range/diameter) ≈ 0.25.

The sensitivity of such transducers depends, of course, in the target material or on its conducting surface. The better the conducting target (or surface), the greater the sensitivity. Due to this it is crucial to carry out a calibration of the eddy current transducer depending on the application in which it is going to be integrated. These transducers present sensitivity around 4 V/mm for aluminium targets. The sensitivity decreases for magnetic materials targets. Typical value for the frequency response is 20 kHz. The linearity (or non-linearity error) presents a typical value of ± 0.5 %.

This is a non-contact transducer. They are normally sold without the target as this is usually the moving metallic part of a machine.

Chapter 4

Digital Encoders

Encoders are digital displacement transducers which typically operate by scanning the bar code recorded on the measurement surface. This code can be as simple as equally spaced bars or bands– for *incremental encoders* - or composed by several tracks with different spacing between the bars – in the case of *absolute encoders*.

Although there are many techniques used in displacement measurement, the inherently easy connection of encoders to digital systems makes them quite popular. They are used in the measurement of linear and angular displacement. In both cases the encoders can be either of absolute or incremental type, although in practice absolute linear encoders are rarely used given the associated cost. As with all devices involving a numerical conversion of a quantity, this quantity is quantified. Therefore only a limited number of positions can be distinguished and, consequently, resolution is always finite.

4.1. Absolute Encoders

4.1.1. Assembly and Operating Principle

The absolute encoders consist of several tracks arranged in parallel on a ruler for linear displacement or in a concentric layout on a disk for angular displacement, each track being divided into N elements of equal size in which the binary word associated with a particular position is materialized according to a certain code, Figure 4.1.

The number of tracks defines the number of bits of each word and these are materialized using complementary physical states (opaque, transparent) to distinguish between logic levels '0 'and '1'. To read the generated code, a light source (e.g., a light-emitting diode - LED) and a light sensor (e.g., a phototransistor) set should be associated with each track, Figure 4.2.

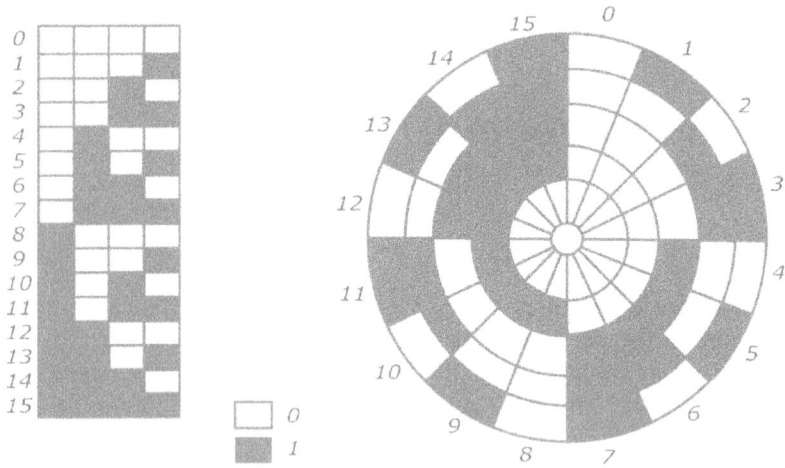

Figure 4.1. Ruler and disk used in 4-bit absolute encoders.

Figure 4.2. Optical encoder disk and light source / light sensor set.

The *n* outputs, corresponding to the *n* bits, are read simultaneously in order to produce a binary representation of position, Figure 4.3.

Each position corresponds to a unique pattern of bits in the various tracks, giving rise to an absolute displacement transducer. Thus, the position is always known and it is not necessary to define a reference position whenever the system is powered down or is turned off.

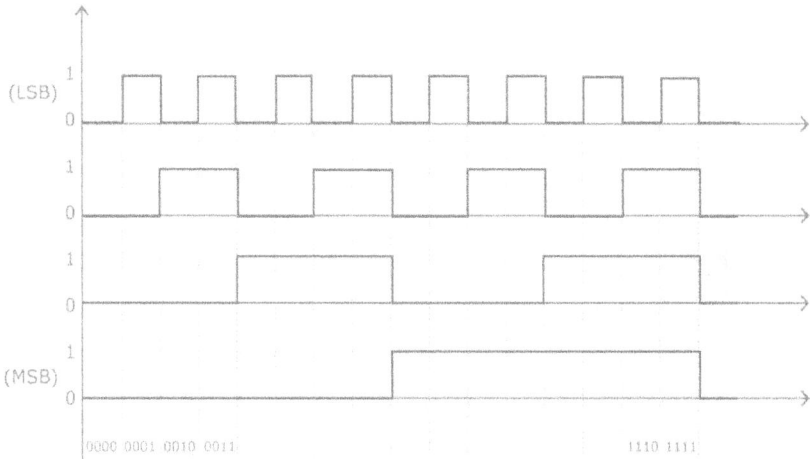

Figure 4.3. Output signals of a 4-bit absolute encoder
(MSB: Most Significant Bit, LSB: Least Significant Bit).

4.1.2. Used Codes

The natural binary code represented in the previous figures has the advantage of being directly used by digital devices. However, when using natural binary code several bits may change simultaneously in the transition from one position to the next. Indeed, if we consider e.g. the transition from code '0111', corresponding to number 7, to code '1000', corresponding to number 8, there is a change of all the bits.

Small misalignments and variations in the response times of different sensor devices (preventing bit changes to be read simultaneously) and the difference of transition time from '1 'to '0' and '0 'to '1' will give rise to spurious codes and hence measurement errors.

These measurement errors can be avoided by using a reflected Gray code. This ensures that consecutive position numbers differ by exactly one bit. Figure 4.4 shows a ruler and a disk implementing this code.

If Gray code output is not compatible with the reading device, a conversion from Gray code to natural binary code may be required. But such conversion is easily accomplished using logic gates. If we consider the n-bit Gray code word, $G_{n-1} G_{n-2} ... G_0$, the corresponding natural binary code word, $B_{n-1} B_{n-2} ... B_0$, can be obtained as follows:

$$B_{n-1} = G_{n-1}$$

$$B_i \mid_{i=0}^{n-2} = B_{i+1} \oplus G_i$$

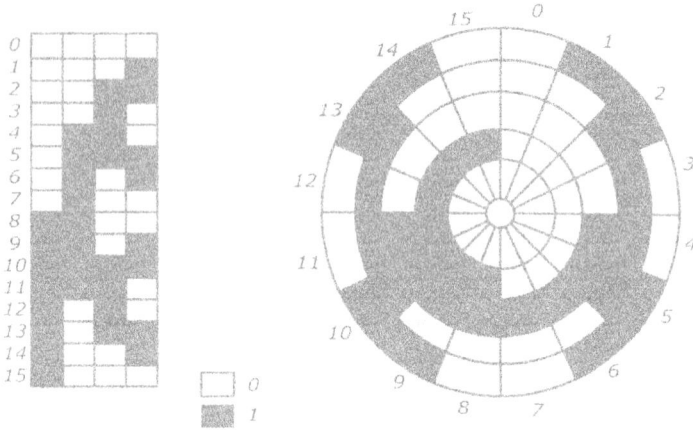

Figure 4.4. Ruler and disk with Gray code.

4.1.3. Some Characteristics of Absolute Encoders

The most important parameters to be considered when specifying an absolute encoder are the resolution, the reading accuracy and the output update maximum frequency.

The number of elements of each track, N, sets the resolution of the encoder. This is given by:

$$\frac{L}{N}, \text{ for a ruler of length } L \text{ (linear encoder)}$$

or

$$\frac{360°}{N}, \text{ for a disk (angular encoder)}$$

As the number of elements of each track is directly related to the number of the encoder bits (e.g. to a 10-bit encoder corresponds

$N = 2^{10}$), resolution is defined by some manufacturers as being n bits. The larger the bits number the better the resolution. Encoders up to 16 bits are the most common, however, encoders with higher resolution may be found.

The resolution of absolute linear encoders depends not only on the number of elements of each track (or on the number of bits) but also on the length of the ruler. For a given number of bits, it is necessary to use different masks for rulers of different lengths. This leads to an increased cost for this type of encoder, which makes it not much used in practice.

Moreover, high resolution does not lead necessarily to good accuracy. The reading accuracy (the maximum error in reading, i.e., the ratio between the measured and the actual position) is directly related to the quality of the coded disk (existence of any imperfection in the mask), to the alignment (or misalignment) between the disk and the reading device and even to the correct assembly of the encoder.

The most relevant characteristics of absolute encoders are now briefly presented.

Strong points:

- Ease of use
- Easy of interface with digital systems
- Good accuracy
- Absolute measurement of displacement (without needing a reference position)

Weak points:

- Bad performance/cost relation (high resolutions encoders are very expensive)

4.2. Incremental Encoders

4.2.1. Assembly and Operating Principle

The incremental encoder consists, in its most basic form, of a disk-shaped track (rotary encoder) or ruler (linear encoder) divided into

$2N$ equal elements alternately opaque and transparent. The reading of the generated code is provided by a light source (e.g., a light emitting diode - LED) and a light sensor (e.g., a phototransistor) placed on either side of the disk, Figure 4.5, or ruler.

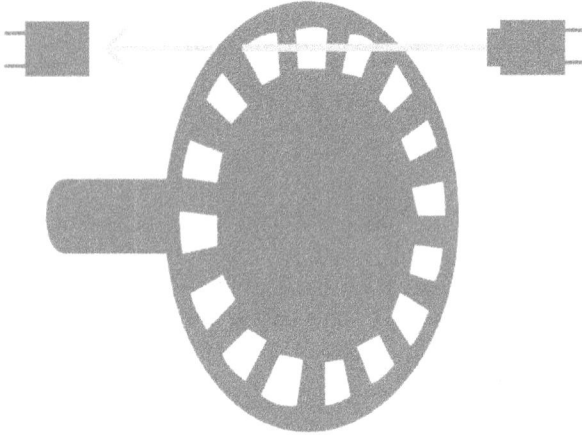

Figure 4.5. Single track incremental rotary encoder.

The disk or ruler can also be made of alternate reflective and non-reflective elements. In this case both the light source and the light sensor are on the same side of the track; the light beam is reflected by the reflective elements and detected by the light sensor.

The disk or ruler movement makes the alternate opaque and transparent elements (or reflective and non-reflective), prevent or allow the arrival of the light beam from the light source to the light sensor. The circuit connected to the light sensor converts this information into an electrical signal. A rectangular wave is then generated with a 'high' logic level when the light reaches the sensor and a 'low' logic level when the light beam is blocked, or vice versa, Figure 4.6.

Figure 4.6. Incremental encoder output signal.

The displacement, relatively to an arbitrary and known origin, is obtained by counting the pulses of a given logic level (either 'high' or 'low') by an electronic counter. The largest measurable displacement will be limited by the number of bits of the counter.

Such incremental encoders allow displacement measurement only in a given direction, for either translation or rotation. If there is a change of direction, the generated pulses are identical, the provided information corresponds to the covered distance and so there will be an error in the displacement measurement.

To overcome this problem, incremental encoders are typically composed of two concentric (disks), Figure 4.7, or parallel (rulers) tracks, with a phase difference of 90° (1/4 of the period), divided into $2N$ equal elements alternately opaque and transparent, each track being associated with a light source / light sensor set, called the A and B channels.

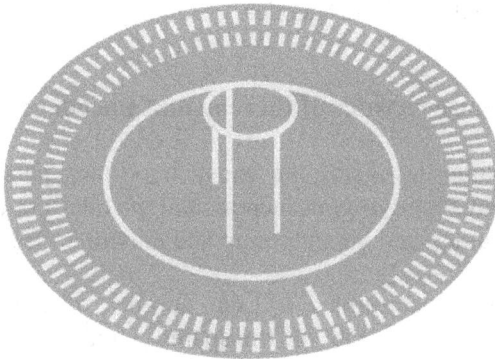

Figure 4.7. Incremental rotary encoder: two track disk.

The obtained signals, Figure 4.8, provide information about the module and also about the direction of displacement. The direction information is obtained by monitoring the channel A and B signals. In one displacement direction, the channel A signal is in advance relative to the channel B signal, in the other direction the opposite occurs.

In this case, the up/down counter increments or decrements the count according to the displacement direction. Using this type of structure the initial period is subdivided into four parts, resulting in a four times higher resolution. Indeed, the signals of channels A and B have four possible states, so it is possible to distinguish, using appropriate electronics, up to four different positions per period.

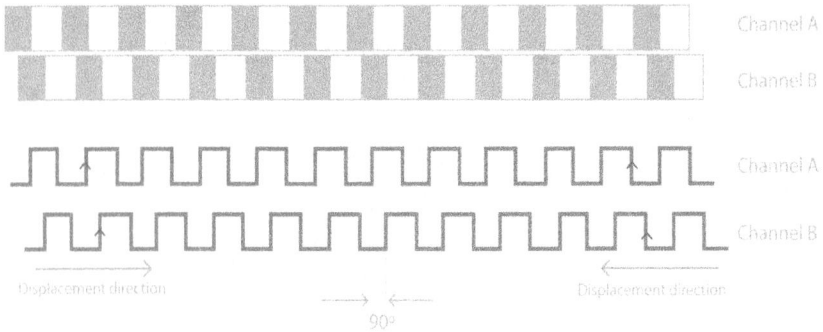

Figure 4.8. Incremental encoder quadrature outputs.

In some cases there is also a third track and corresponding reading device, which gives only one pulse. This can be useful to count the number of revolutions (for angular encoders) and / or for determining the reference position of the encoder.

4.2.2. Some Characteristics of Incremental Encoders

As with absolute encoders, the most important parameters to be considered when specifying an incremental encoder are the resolution, the reading accuracy and the output update maximum frequency.

The resolution is determined by the number of the disc or ruler elements, N, either opaque or transparent, and the used decoding method.

The most relevant characteristics of incremental encoders are now briefly presented.

Strong points:

- Ease of use
- Ease of interface with digital systems
- Good performance/cost relation
- Good accuracy

Weak points:

- Requires a well-known position reference

- Pulses resulting from electrical noise give rise to errors which persist even when the noise disappears

- The displacement information is lost whenever the system is powered down or is turned off. However, particularly in robotic systems or machine tools the encoder / counter set is continuously powered by batteries.

Chapter 5

Fibre Optic Displacement Sensors[1]

5.1. Introduction

Optical fibre based sensor technology offers the possibility of developing a variety of physical sensors for a wide range of physical parameters. Fibre sensors can be designed so that the measurand interacts with one or several optical parameters of the guided light (intensity, phase, polarization and wavelength). In general, the main interest in this type of sensors comes from the fact the optical fibre itself offers numerous operational benefits. It is electromagnetically passive, so it can operate in high and variable electric field environments (like those typical of the electric power industry) and where there is explosion risk; it is chemically and biologically inert since the basic transduction material (silica) is resistant to most chemical and biological agents; its packaging can be physically small and lightweight; due to the intrinsic low optical attenuation of the fibre, it is possible to attain distributed sensing, i.e. determine the measurand as a function of the position along the length of the fibre; the optical fibre can be operated over very long transmission lengths, so the sensor can easily be placed kilometers away from the monitoring station; finally, it is also possible to perform multiplexed measurements using large arrays of remote sensors, operated from a single optical source and detection unit, with no active optoelectronic components located in the measurement area, thereby retaining electromagnetic passiveness and environmental resistance.

This chapter aims to provide some insights on fibre optic based displacement sensors. Section 5.2 addresses a general classification of contact and non-contact fibre optic displacement sensors, in each case presenting some specific but important configurations. Section 5.3 details the characteristics of selected sensing layouts from the

---

[1] José Luís Santos, Luís Alberto Ferreira, Francisco Moita Araújo (invited authors).

development of simple models, illustrating methodologies to study this type of sensors. Section 5.4 elaborates on the displacement resolution limits associated with optical sensing in general and some insights are delivered on conceptual new developments. Finally, concluding remarks are presented in section 5.5.

5.2. Classification of Fibre Optic Displacement Sensors

A displacement sensor measures the distance that an object moves. It can be of two types, one that involves a contact with the object and the other where such contact is not needed, in this case always involving the interaction of the object with a field generated in the sensor structure (optical, acoustic, magnetic, etc.). Optical fibre based displacement sensors rely on the modulation of one of the light properties through some kind of mechanism coupled to the object displacement. This mechanism defines an operation mode which implicates either contact or non-contact with the optical fibre.

Figure 5.1 (a) shows a general scheme for the first type of sensors in which the light does not exit the optical fibre in the sensing region (therefore, integrated in the general class of optical fibre intrinsic sensors). The mechanism coupled to the moving object introduces a mechanical action on the fibre which changes the properties of the light that propagates in the fibre. For example, as shown in Figure 5.1 (b), the sensing region may consist of two corrugated plates. The corrugations are cylindrical rods with fixed diameters. The fibre is pressed between these plates due to the object displacement, therefore introducing microbending loss. In this situation it is the light intensity that is modulated by the displacement, which means that we deal with an optical fibre intensity sensor. However, other modulation alternatives are possible. For example, the displacement can induce optical fibre path variations in the sensing region, which can be read with high sensitivity introducing such section of fibre into one arm of a Mach-Zehnder interferometer. In this case we have an interferometric optical fibre displacement sensor.

Figure 5.1 (c) illustrates another type of contact fibre optic displacement sensor where is included a spring element to change the displacement sensitivity of the fibre optic element, for example a Fibre Bragg Grating (FBG).

The stiffness of silica is rather high (Young's modulus $E = (7.29 \pm 1.6) \times 10^{10}$ N·m^2, meaning that a substantial axial force N is required to change the length of standard optical fibres (the strain is $0.11\%/N$ for optical fibres with 125 mm external diameter), with the corresponding difficulty to follow the displacement of the object (assuming that the load of the sensing element on the object must be residual). If a spring with a lower stiffness compared with the optical fibre is added in series with the fibre, then the change in the total length L_{Total} will essentially appear on the spring, permitting larger displacements, not possible by considering uniquely the fibre sensing element. The characteristics of the displacement sensor shown in Figure 5.1(c) will be detailed later.

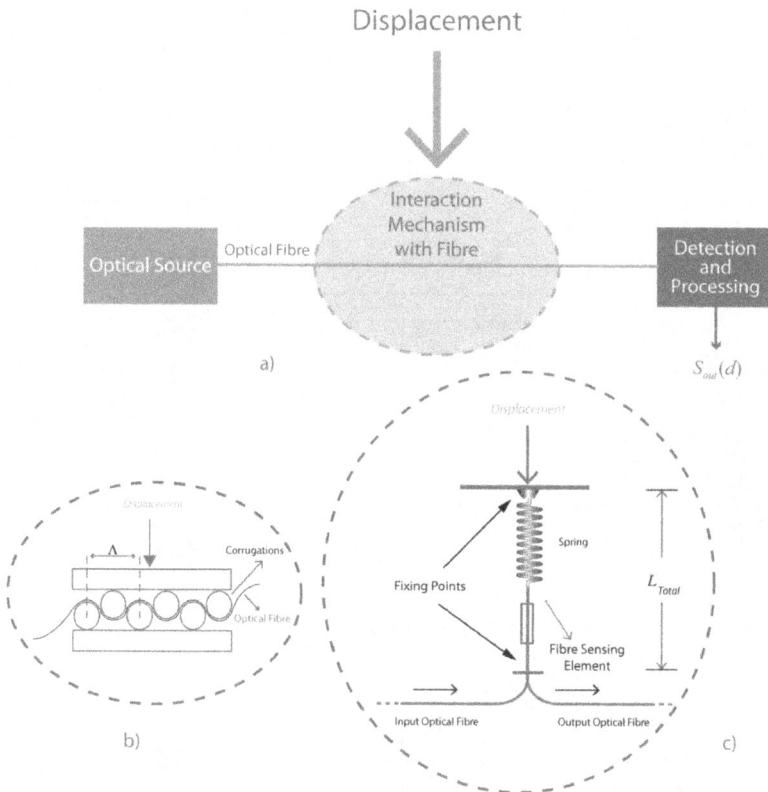

Figure 5.1. General scheme of a contact fibre optic displacement sensor (a) and two layouts of the sensing head, with a microbend structure (b) and with a configuration involving the coupling to a spring element to increase the displacement range (c).

Non-contact optical fibre displacement sensors require the light to exit the fibre and, therefore, are in the group of optical fibre extrinsic sensors. Figure 5.2 shows some possible configurations.

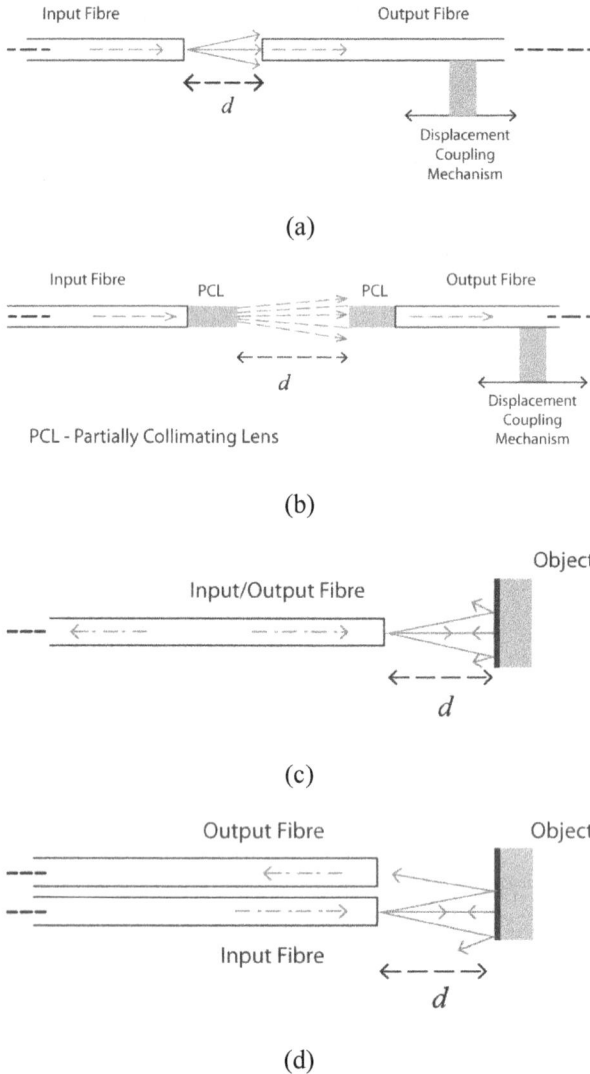

(a)

(b)

(c)

(d)

Figure 5.2 (a-d). Some examples of non-contact optical fibre displacement sensors. The configurations from (a) to (d) are associated with displacement induced intensity modulation, while in configurations (e) and (f) it is the light optical phase which is modulated by the displacement.

(e)

(f)

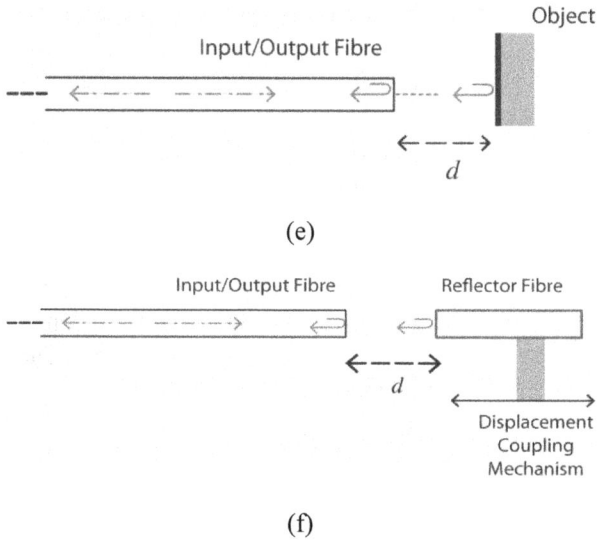

Figure 5.2 (e-f). Some examples of non-contact optical fibre displacement sensors. The configurations from (a) to (d) are associated with displacement induced intensity modulation, while in configurations (e) and (f) it is the light optical phase which is modulated by the displacement.

Figure 5.2(a) illustrates the situation in which the light exits the input fibre (typically a multimode fibre with a core and external diameters of 100 µm and 125 µm, respectively) and propagates in the air along a cone, with an angle θ_{max} in relation to the fibre axis, determined by the fibre numerical aperture, NA. Typically $NA = 0.25$, giving $\theta_{max} = sin^{-1}(0.25) \approx 14.5°$. The output fibre is located at a distance d from the input fibre, therefore only a fraction of the light that exits this fibre is coupled to the output fibre. This fibre is attached to a mechanism which couples to the object whose displacement is to be measured. When the object moves, the distance d changes and the optical power, coupled to the second fibre also changes, establishing a dependence that permits to determine the object displacement by monitoring the output optical power.

Figure 5.2(b) shows the same transducing principle, but now the diverging angle of the light that propagates in the air is reduced by means of a partially collimating lens. In this way the measurement range can be adjusted by changing the collimation degree, albeit at the expense of a smaller sensitivity of the output optical power variations to changes in d.

Figures 5.2 (a) and 5.2 (b) illustrate transmissive layouts but reflective ones are also possible and indeed more common essentially because they are simpler to assemble. Figure 5.2 (c) shows one of these configurations where the input fibre is also the output fibre. The light exiting the fibre is reflected on a mirrored surface attached to the object whose movement is to be monitored. A fraction of the reflected light is reinjected into the fibre, with a value that depends on distance d. The advantage of this scheme is its simplicity, but the alternative structure with two fibres shown in Figure 5.2 (d) is also popular. Here, the input and output fibres are distinct, adding extra flexibility in tailoring the dependence of the intensity in the return fibre on distance d, for example by changing the lateral separation between the two fibres.

The sensing layouts shown in Figures 5.2 (a), (b), (c), (d) rely on displacement induced intensity modulation of the light. Therefore, they belong to the class of fibre optic intensity sensors. However, other type of modulation is possible. For example, Figures 5.2 (e) and 5.2 (f) shows interferometric sensing structures in the sense that the movement of the object changes the phase of the reflected light reinjected into the input fibre (signal beam), which interferes with the reference beam generated by the Fresnel reflection in the fibre tip. Therefore, we are in presence of a Fabry-Pérot interferometric cavity, with a transfer function given by:

$$I_{out} = I_0 \left[1 + k \cdot \cos\phi \right] \qquad (5.1)$$

where I_{out} is the optical power that propagates back into the input/output fibre to the detector, I_0 is proportional to the optical power injected from the source into the input fibre, k is the fringe visibility and the phase ϕ is given by:

$$\phi = \frac{4 \cdot \pi \cdot d}{\lambda} \qquad (5.2)$$

where λ is the wavelength of the light. These two relations indicate that I_{out} changes substantially for variations of d which are a fraction of the wavelength (typically of the order of 1 μm), indicating how sensitive is the interferometric readout of the object movement.

In Figure 5.2 (e), the light is reflected by a mirrored surface attached to the object, while in Figure 5.2 (f) the light is reflected on the tip of a fibre mechanically fixed to the object. Contrary to the situations where light intensity modulation was involved, now the input/output fibre needs to be single mode (core diameter of ≈ 10 μm) to preserve the spatial coherence of the light emitted by a coherent light source, such as a laser. Therefore, interferometric sensing structures are substantially more demanding in what concerns the technology involved compared with intensity based sensing.

The sensing configurations of the type illustrated in Figures 5.1 and 5.2 require two conditions for their implementation: the optical fibre must be close the object whose displacement is to be measured; displacement variations should be relatively small. In many situations one of these conditions, or both, are not fulfilled and another approach is needed. In general, this involves the generation of a carrier with a frequency well below the optical frequency, which requires the modulation of one parameter of the light emitted by the optical source. Two methodologies have been followed. One involves the modulation of the frequency of the emitted light in a saw tooth format and the optical interference of two beams, one reflected by the object and the other generated in a fixed position and working as a reference beam. The time delay between these two beams, dependent on the initial set-up and the displacement of the object, is translated into a beat frequency, small enough to be detected and transformed into an electrical carrier. The displacement information is, therefore, obtained by monitoring the value of this beat frequency.

The second methodology is based on the sine wave modulation of the optical power emitted by the source. The two optical sine waves (signal and reference) are converted into electric sine waves after photo detection, and their combination results into a sine wave signal with a phase that is a function of the time delay between the two optical beams, again dependent on the object displacement. This phase can be recovered using lock-in techniques and the distance to the object determined after proper calibration. Figure 5.3 illustrates a typical layout for the implementation of these two sensing methodologies, showing also that the collimation of the signal light allows the object to be far away from the sensing lead fibre, overcoming one of the limitations indicated above.

In order to explain in more detail the main characteristics of optical fibre displacement sensors, one configuration from each of the three kinds of sensors illustrated in Figures 5.1, 5.2 and 5.3 was selected for further analysis, which is presented in the next section.

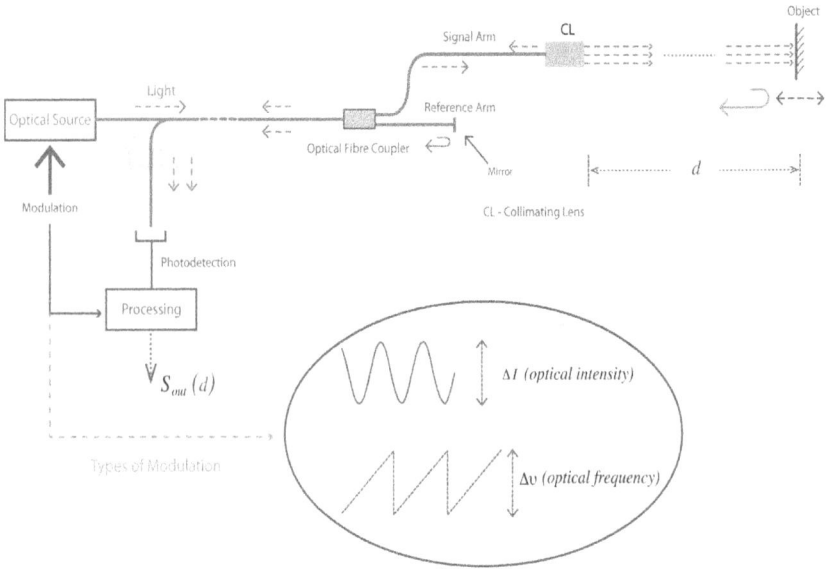

Figure 5.3. Illustrative layout of fibre optic displacement sensors based on sub-carrier generation, either by sine wave modulation of the emitted source optical power or by saw tooth modulation of the optical frequency of the light emitted by the source.

5.3. Analysis of Specific Sensing Configurations

In this section three specific configurations will be explored: the structure shown in Figure 5.1(c) as an example of a contact type displacement sensor with an FBG sensing element; the layout indicated in Figure 5.2(d) illustrative of a non-contact displacement sensor; and the configuration presented in Figure 5.3.

5.3.1. Spring Interfaced Fibre Optic Displacement Sensor

In this section the sensing structure shown in Figure 5.1(c) will be further analyzed. By proper adjustment of the constants of the spring, k_s, and of the fibre, k_f, the sensor can be tailored to measure a specific displacement range. Figure 5.4 illustrates the relevant parameters for this analysis, as well as the specification of the sensing element considered here, which is a FBG.

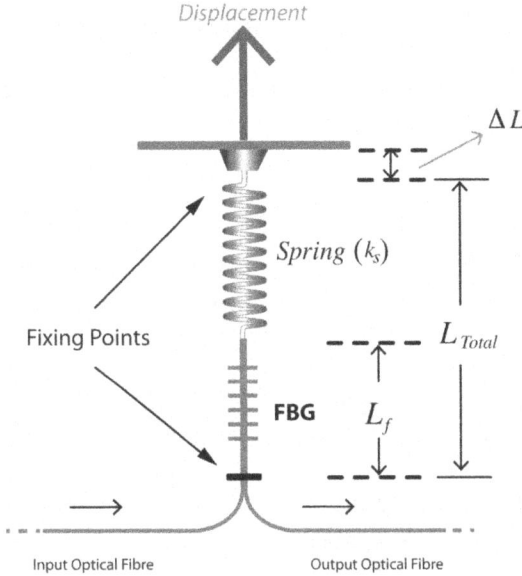

Figure 5.4. Schematic of a contact fibre optic displacement sensor where the sensing element is a fibre Bragg grating with the displacement range adjusted by proper selection of the spring constant K_s.

As shown in the Figure 5.4, L_{Total} is the distance between the fixing points with a part in fibre (length L_f). The displacement ΔL appears as a sum of two components, one as the elongation of the spring (ΔL_s) and the other as that of the fibre (ΔL_f), i.e.,

$$\Delta L = \Delta L_s + \Delta L_f \qquad (5.3)$$

The length increase of the spring and fibre originates restoring forces given in magnitude by $F_s = k_s \Delta L_s$ and $F_f = k_f \Delta L_f$, respectively. The equilibrium of the system requires

$$F_s = F_f \Rightarrow \Delta L_s = \frac{k_f}{k_s} \Delta L_f \qquad (5.4)$$

Combining (5.3) and (5.4) leads to

$$\Delta L = \Delta L_f \left(\frac{k_s + k_f}{k_s} \right) = \frac{\Delta L_f}{L_f} L_f \left(\frac{k_s + k_f}{k_s} \right) = \Delta \varepsilon_f L_f \left(\frac{k_s + k_f}{k_s} \right), \qquad (5.5)$$

where $\Delta\varepsilon_f$ is the change in fibre strain due to the displacement ΔL.

If the sensing element is a FBG, the change of its Bragg wavelength, λ_B, due to $\Delta\varepsilon_f$ is given by:

$$\Delta\lambda_B = k_{FBG}\Delta\varepsilon_f \,, \tag{5.6}$$

where $k_{FBG}=\lambda_B(1-p_\varepsilon)$, with p_ε the photo-elastic constant. For a typical fibre $p_\varepsilon \approx 0.22$. The combination of (5.5) and (5.6) gives

$$\Delta L = \frac{L_f}{k_{FBG}}\left(\frac{k_s + k_f}{k_s}\right)\Delta\lambda_B \tag{5.7}$$

An expression for k_f may be obtained by using Hooke's law to relate changes in stress (ΔT) and strain:

$$\Delta T \equiv \frac{F_f}{A_f} = E_f\,\Delta\varepsilon_f \tag{5.8}$$

with A_f the area of the fibre cross section and E_f the fibre Young modulus. Considering $F_f = k_f\Delta L_f$, then

$$A_f E_f\,\frac{\Delta L_f}{L_f} = k_f\,\Delta L_f \Rightarrow k_f = \frac{A_f}{L_f}E_f \tag{5.9}$$

For a standard fibre, the diameter is 125 μm and $A_f=1.23\times10^{-8}$ m^2. On the other hand, the Young modulus is $E_f \approx 7.3\times10^{10}$ Nm^{-2}. Therefore, $k_f \approx 8.98\times10^2(1/L_f)$ N.

When $k_f \gg k_s$ equation (5.7) becomes

$$\Delta L \approx \frac{L_f}{k_{FBG}}\left(\frac{k_f}{k_s}\right)\Delta\lambda_B \tag{5.10}$$

which shows that for a certain Bragg wavelength shift, $\Delta\lambda_B$, the associated value for ΔL is much larger than the one that originates the same $\Delta\lambda_B$ when no spring is considered, illustrating the extension of the displacement measurement range when the sensing structure shown in

70

Figure 5.4 is considered (certainly at the expense of a reduced sensitivity).

5.3.2. Reflection Fibre Optic Displacement Sensor

This type of sensors with physical operation based on the dependence of the returned optical power on the distance from the fibre(s) to the object has been widely used for long, essentially due to its simplicity and reliability. Here we are going to analyze the configuration with two fibres, shown in Figure 5.2 (d). A more detailed layout is given in Figure 5.5.

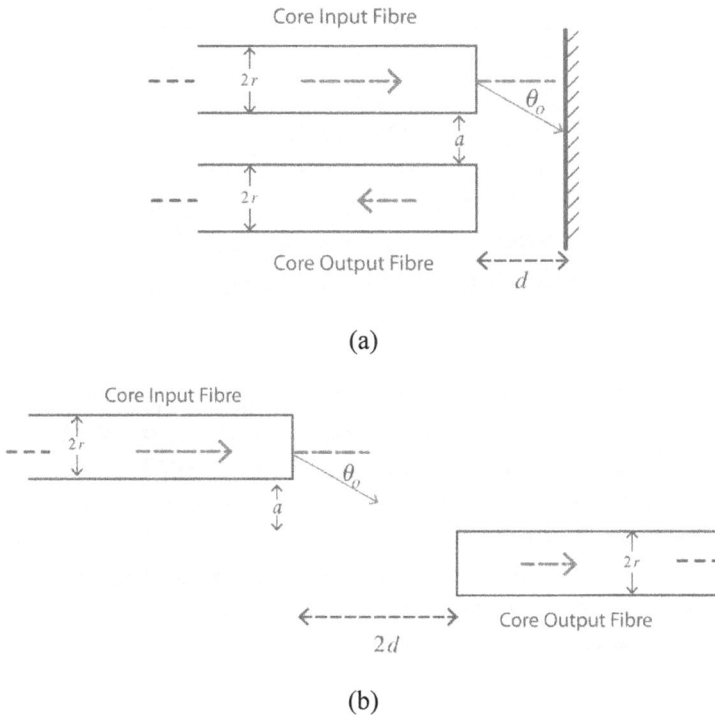

(a)

(b)

Figure 5.5. Reflection sensor with two fibres (a). For analysis purposes, layout (b) is equivalent to layout (a).

What is represented in Figure 5.5 (a) is the core of a fibre (the cladding is not drawn), with radius r. The cores are separated by distance a. The problem consists in determining the power coupling between the two fibres as a function of a certain number of parameters, particularly d. From the analysis point of view, scheme (a) is equivalent to (b), where the fibres are axially separated by a and longitudinally by $2d$. The analysis that follows is purely based in geometric optics, an acceptable option given that multimode fibres are considered.

The analysis of Figure 5.5 indicates the need to consider three cases, i.e.,

$$d < \frac{a}{2T} \Rightarrow \text{the return fibre is not illuminated}$$

$$\frac{a}{2T} \le d < \frac{(a+2r)}{2T} \Rightarrow \text{a fraction of the input face of this fibre is illuminated}$$

$$d \ge \frac{(a+2r)}{2T} \Rightarrow \text{the input face of the return fibre is fully illuminated}$$

(5.11)

where

$$T = \tan(\arcsin NA); \quad NA \equiv \sin\theta_0 \text{ is the fibre numerical aperture} \quad (5.12)$$

A simple approach consists in considering the cross section of the radiation cone exiting the fibre with a uniform power density. In these circumstances, the power efficiency of the sensor (ratio of the optical power that propagates down the return fibre and the illumination optical power) is given simply as

$$\eta = \left(\frac{r}{2T}\right)^2 \tag{5.13}$$

However, this is a rather crude model and for its improvement it is necessary to consider the non-uniform distribution of the optical power in the radiation cone, which is given by:

$$p(\theta) = p(0)\left[1 - \frac{\sin^2\theta}{NA^2}\right]^{2/\alpha}, \tag{5.14}$$

where α is a parameter that defines the radial profile of the core refractive index (when $\alpha \to \infty$ the fibre becomes step-index, situation in which $p(\theta) = p(0)$). This equation assumes a uniform power distribution through the fibre modes and far-field operation, conditions fairly fulfilled in the situation illustrated in Figure 5.5.

To evaluate the amount of optical power coupled to the return fibre let us define the relevant parameters as shown in Figure 5.6.

The surface element dS in spherical coordinates is

$$dS = 4d^2 \cdot \sin\theta \cdot d\theta \cdot d\phi \qquad (5.15)$$

The optical power dP that reaches dS is given by

$$dP = \frac{p(\theta)}{2\pi} d\Gamma \qquad (5.16)$$

where $d\Gamma$ is the solid angle element due to dS seen from O. Therefore, it is given by $d\Gamma = dS/4d^2$. The total optical power in the radiation cone is calculated from

$$P_{cone} = \frac{p(0)}{2\pi} \int_{0}^{2\pi} \int_{0}^{\theta_0} \left[1 - \left(\frac{\sin\theta}{NA} \right)^2 \right]^{\frac{2}{\alpha}} \sin\theta \cdot d\theta \cdot d\varphi \qquad (5.17)$$

On the other hand, the optical power intercepted by the return fibre is

$$P_{int} = \frac{p(0)}{2\pi} \int_{0}^{\phi_0} \int_{\theta_1}^{\theta_2} \left[1 - \left(\frac{\sin\theta}{NA} \right)^2 \right]^{\frac{2}{\alpha}} \sin\theta \cdot d\theta \cdot d\phi \qquad (5.18)$$

with the angles illustrated in Figure 5.6. The objective is to integrate on the area of the fibre tip as seen from O. In this situation the fibre end takes an elliptical shape with major and minor semi-axes given by r and $r\cos\theta$, respectively. In fact, considering $\theta \leq \theta_0$ and $\cos\theta_0 = 0.96$ when $NA = 0.28$ the ellipse does not deviate substantially from a circle of radius r. Indeed, the area of the ellipse is

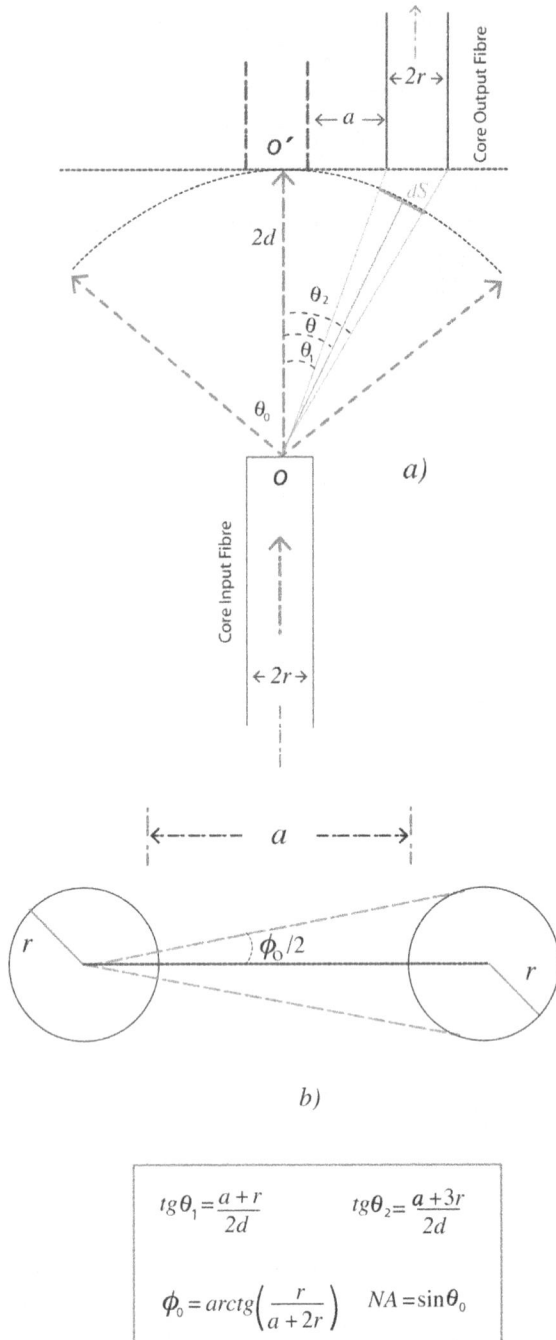

Core Output Fibre

$\leftarrow 2r \rightarrow$

$\leftarrow a \rightarrow$

O'

2d

θ_2

θ

θ_1

θ_0

O

a)

Core Input Fibre

$\leftarrow 2r \rightarrow$

$\leftarrow \quad a \quad \rightarrow$

r

$\phi_0/2$

r

b)

$$tg\,\theta_1 = \frac{a+r}{2d} \qquad tg\,\theta_2 = \frac{a+3r}{2d}$$

$$\phi_0 = arctg\left(\frac{r}{a+2r}\right) \qquad NA = \sin\theta_0$$

Figure 5.6. Geometry for the determination of the coupling efficiency: definition of the polar angles (a) and azimuthal of integration (b).

$$S_{int} = \int\limits_{0}^{\phi_0} \int\limits_{\theta_1}^{\theta_2} 4d^2 \sin\theta \cdot d\theta \cdot d\phi = 8d^2 \text{arctg}\left(\frac{4}{a+2r}\right)(\cos\theta_1 - \cos\theta_2)$$

$$(5.19)$$

Considering $\tan\theta \approx \theta$ and $\cos\theta \approx 1 - \theta^2/2$, we get

$$S_{int} = 4r^2 = 1.27\pi r^2 = 1.27 S_{real} \qquad (5.20)$$

where $S_{real} = \pi r^2$ is the area of the fibre end face. To simplify, the integration limits in equation (5.19) will be kept, with the error associated with the wrong integration area attenuated by including in such equation the numerical factor $\Psi = 1/1.27 = 0.79$.

For step-index fibres the rays exiting the input fibre and incident on the core of the return fibre (after reflected in the mirror) are always guided, a factor that considerably simplifies the analysis. What needs to be considered is the Fresnel reflection on the end face of the return fibre, which will be modelled by the inclusion of a (transmissive) numerical factor F (for the air-silica interface $F \approx 0.96$).

From the definition of the coupling efficiency applied to the present configuration

$$\eta = \frac{P_{ret}}{P_{cone}} \qquad (5.21)$$

with $P_{ret} = \Psi\beta F P_{int}$, where β is the fraction of the illuminated core area of the return fibre, and considering equations (5.17) and (5.18) for P_{cone} and P_{int} we get

$$\eta = F \cdot \Psi \cdot \beta \frac{\text{arctg}\left(\frac{r}{a+2r}\right)(\cos\theta_1 - \cos\theta_2)}{\pi(1-\gamma)} \qquad (5.22)$$

with

$$\cos\theta_1 = \frac{2d}{\sqrt{4d^2 + (a+r)^2}}$$

$$\cos\theta_2 = \frac{2d}{\sqrt{4d^2 + (a+3r)^2}} \qquad (5.23)$$

$$\gamma = \sqrt{1 - NA^2}$$

To evaluate β the analysis is substantially simplified if the intersection of the light cone with the core of the return fibre is linearized, as indicated in Figure 5.7 (b).

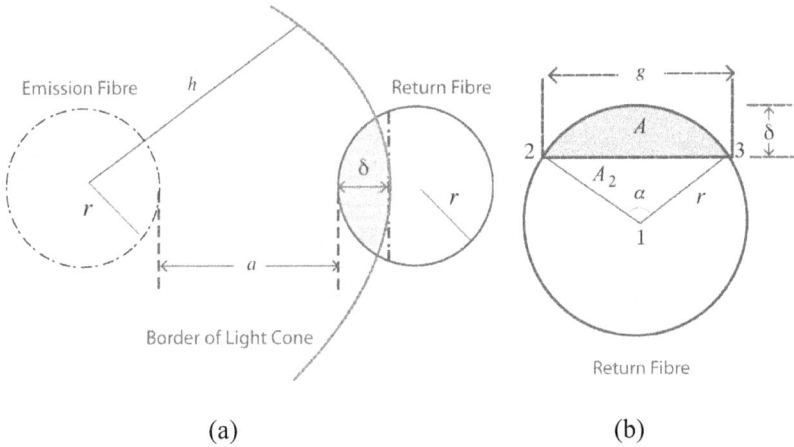

(a) (b)

Figure 5.7. Geometry associated to the partial illumination of the return fibre (a) and simplified geometry layout whereby the intersection of the light cone with the input face of the return fibre is approximated by a line segment (b).

The area A_1 of the circular sector associated with the angle α is given by $A_1 = \frac{1}{2}r^2\alpha$; the area of the triangle $(1,2,3)$ is

$$A_2 = \frac{1}{2}gr\cos\left(\frac{\alpha}{2}\right) = \frac{1}{2}r^2\sin\alpha \qquad (5.24)$$

The intercepted area (dotted part in Figure 5.7b) is

$$A \equiv A_1 - A_2 = \frac{1}{2}r^2\left(\alpha - \sin\alpha\right) \tag{5.25}$$

Considering $r - \delta = r \cdot \cos\left(\dfrac{\alpha}{2}\right)$ and $\alpha = 2\arcos\left(1 - \dfrac{\delta}{r}\right)$, it results

$$A = r^2\left[\arcos\left(1 - \frac{\delta}{r}\right) - \left(1 - \frac{\delta}{r}\right)\sin\left[\arcos\left(1 - \frac{\delta}{r}\right)\right]\right] \tag{5.26}$$

The area of the end face (core) of the return fibre is πr^2. Therefore,

$$\beta = \frac{1}{\pi}\left[\arcos\left(1 - \frac{\delta}{r}\right) - \left(1 - \frac{\delta}{r}\right)\sin\left[\arcos\left(1 - \frac{\delta}{r}\right)\right]\right] \tag{5.27}$$

From Figures 5.6 and 5.7, the parameter δ depends on the distance d and the numerical aperture of the fibre. Observation of these figures permits to obtain

$$\delta = 2dT - a \tag{5.28}$$

where T is given by (5.12). To summarize, for the three cases of equation (5.11) the optical coupling efficiency from the input to the output fibre is:

$$0 < d < \frac{a}{2T}, \eta = 0$$

$$\frac{a}{2T} \leq d < \frac{(a+2r)}{2T}, \eta \text{ given by (5.21) with } \beta \text{ calculated from (5.27)} \tag{5.29}$$

$$d \geq \frac{(a+2r)}{2T}, \eta \text{ given by (5.21) with } \beta = 1$$

The relation (5.27) gives an approximation for the illuminated fraction of the return fibre core, permitting to evaluate with a simple model the behaviour of the transfer function of the fibre optic sensor illustrated in Figure 5.5 a). A more elaborate analysis allows to obtain the exact value for β, showing that the error originated by the application of the simplified model does not exceed 10 % in most situations.

Figure 5.8 shows the variation (in percentage) of the power coupling efficiency η versus the displacement d for two different values of a.

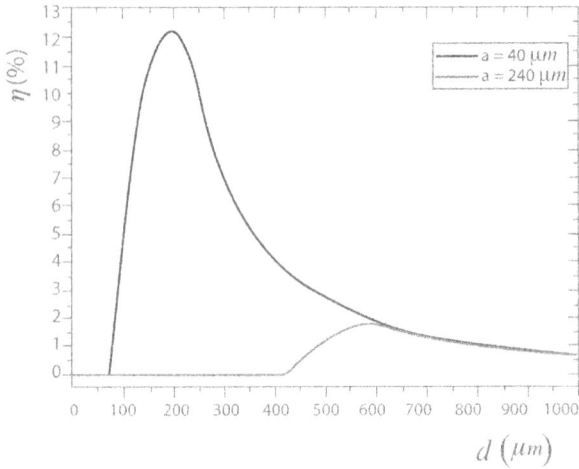

Figure 5.8. Transfer function of the two-fibre reflective fibre optic displacement sensor for two different values of a.

The following features emerge from the observation of Figure 5.8: i) the coupling efficiency increases for smaller values of a (for fibres with standard 125 μm diameter, the minimum value of a equals 25 μm, situation in which the two fibres are in contact); ii) there are two operational regions, one associated with the increase of η with d (largest slope and a fairly linear behaviour) , and the other where η decreases with d (non-uniform slope, i.e., variable sensitivity); iii) the sensitivity can be adjusted by changing the value of a.

The two-fibre displacement sensor is an effective sensing structure, with a flexibility level that permits to tailor the sensor for a particular displacement measurement application. However, this is not the most used reflective fibre sensor configuration, a status that is attributed to the single fibre displacement sensor shown in Figure 5.2c), in which the fibre provides simultaneous illumination and return. The transfer function of this sensor is shown in Figure 5.9 for a multimode fibre with 100 μm core diameter, Figure 5.9a), and a single mode fibre (core diameter ≈ 10 μm; Figure 5.9 b)).

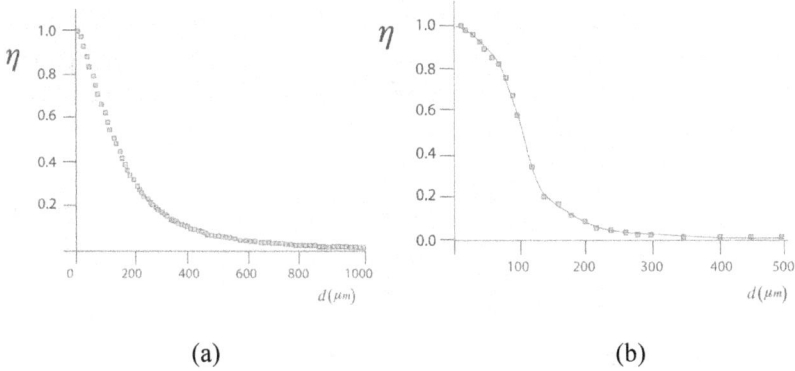

(a)   (b)

Figure 5.9. Transfer function of a single-fibre displacement sensor: for a 100 μm core diameter multimode fibre (a) and for a 10 μm core diameter single mode fibre (b).

It can be observed that for the 100 μm core diameter multimode fibre the decrease of η with d is faster when compared with that of the 10 μm core diameter single mode fibre, the value $\eta = 0.5$ being reached for $d \approx 150$ μm and $d \approx 120$ μm, respectively.

The optical sensing approaches based on the measurand induced modulation of the optical power has the problem of distinguishing the power changes associated with the sensor operation from those that always appear in optical systems, associated for example with non-constant optical source light emission, variable losses in components, variable gain in the photo detection process and so on. So, for these sensors to be effective adequate referencing techniques need to be coupled to any specific sensing structure. The same applies to fibre optic reflective displacement sensors, and several schemes have been developed to implement optical power referentiation. For the case of a single fibre displacement sensor, Figure 5.10 shows an effective technique to achieve such functionality, based on the integration of a fibre Bragg grating (FBG) in the sensing head.

In Figure 5.10, the optical power that propagates down to the sensing head is P_{in}. A slice of this power at the Bragg wavelength λ_B of the FBG is reflected by this device, generating P_{ref}, the referencing component, and the remaining optical power, P_{obg}, propagates to the object whose displacement is to be measured. After reflection, the recoupled optical power modulated by distance d is $P_{ret}(d)$, that propagates down the fibre to the photo detection unit (the FBG being transparent to this optical power because it does not have the power slice centered at λ_B). In this unit a spectral component spatially separates P_{ref} from $P_{ret}(d)$, generating after detection the voltage signals V_{ref} and $V_{ret}(d)$. With the processing

$$S_{out}(d) \equiv \frac{V_{ret}(d)}{V_{ref}} \qquad (5.30)$$

the signal $S_{out}(d)$ is generated which is independent of the optical power source fluctuations and variable losses that can occur along the optical system, providing a referenced signal for the determination of the displacement d.

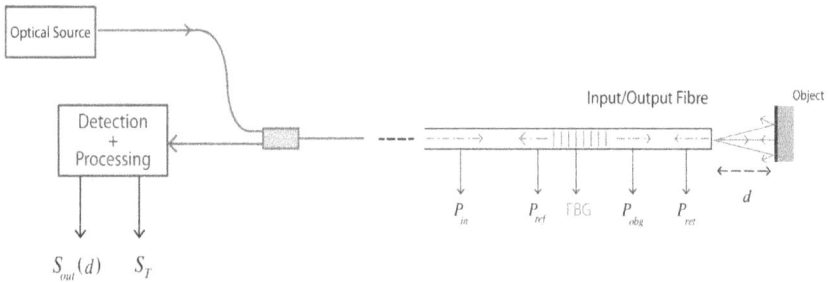

Figure 5.10. Schematics of a single-fibre displacement sensor incorporating a referencing technique based on the use of a FBG.

An additional bonus of the configuration indicated in Figure 5.10 is the determination of the temperature, T_{sh}, of the sensing head. The wavelength, λ_B, of the light reflected by the FBG is temperature dependent. Therefore, with adequate processing, it is possible to generate the signal S_T in Figure 5.10, which is proportional to T_{sh}.

5.3.3. Fibre Optic Displacement Sensor Based on Sub-carrier Interferometry

As indicated in the context of Figure 5.3, the application of techniques based on the generation of a sub-carrier, with a frequency well below the optical frequency, permit to detect large displacement of objects located far away from the end face of the optical fibre that guides the light from the optical source. So, these approaches are complementary

80

to the ones illustrated in Figures 5.1 and 5.2. One of these techniques is outlined here.

One of the most well-known source modulation interferometry techniques is the *Frequency Modulation Continuous Wave* (FMCW). It is based on the saw tooth modulation of the source optical frequency. As indicated in Figure 5.3, two waves are generated that optically interfere before photo detection: a reference wave, returned from the reference arm, and a signal wave that is returned from the object whose displacement is to be measured and which accumulates a time delay associated with the distance to the object. The identification of the relevant parameters is shown in Figure 5.11.

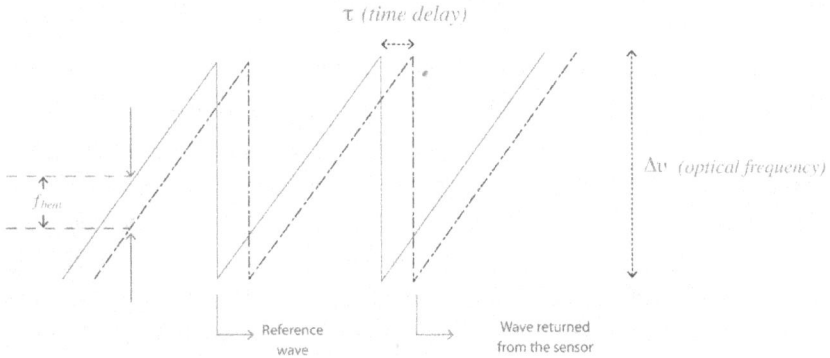

Figure 5.11. Interference waves in the Frequency Modulation Continuous Wave technique.

The amplitude of the serrodyne source optical frequency modulation is $\Delta \upsilon$ and the time delay is τ. When the two waves mix coherently, the beat frequency f_{beat} is dependent on the delay but also on the rate of change, $d\upsilon/dt$, of the absolute frequency of the optical source. As the scanning range is finite, two beat frequencies are generated, the lower one (f_{beat}) and the upper one (that appears when one of the waves in Figure 5.11 already did the flyback, while the other keeps in the previous period). The upper one is typically quite high, not detectable by the photo detection and amplification block due to its limited bandwidth. The lower beat frequency is given by

$$f_{beat} = \frac{d\upsilon}{dt}\tau \qquad (5.31)$$

Considering $\tau = \tau_0 + \dfrac{2d}{c}$, where τ_0 is the constant associated with different propagation paths in fibre of the two waves, we get

$$f_{\text{beat}} = f_0 + \frac{2}{c}\left(\frac{d\upsilon}{dt}\right)d \qquad (5.32)$$

where f_0 is the constant frequency due to τ_0 and c the speed of light in air. This equation shows that $(f_{\text{beat}} - f_0)$ is proportional to d, i.e., a distance is obtained by measuring a frequency of an harmonic electric signal, which can be done with high precision and does not need any type of optical power referentiation.

The optical source is typically a laser diode, with standard values for $d\upsilon/dt$ between 10^{13} and 10^{14} Hz/s. Considering $d\upsilon/dt = 10^{14}$ Hz/s, the beat frequency changes equal 3.3 kHz for a path imbalance variation of $\Delta d = 1\,\text{cm}$, and 0.3 Hz for imbalance variation of $\Delta d = 1\,\mu\text{m}$. This value of frequency change is possible to be detected, indicating the level of minimum displacement variation that can be measured with a standard implementation of this technique. On the upper level, displacement variations of several metres can be measured with this approach, the limitation being associated with the coherence length of the optical source. Anyway, for larger displacements other techniques are more adequate than the FMCW. Besides the application of standard optical radar techniques, one of them is based not on the optical modulation of the source but, instead, on the modulation of its intensity in a sine wave format (also illustrated in Figure 5.3). The reference wave is not generated in the reference arm but is derived directly from the signal generator, which is coupled into the reference channel of a lock-in amplifier. After photo detection and amplification, the signal sine wave is coupled into the signal channel of the lock-in. After removal of a constant term, the phase difference between these two waves is proportional to the distance d of the object. This technique can be implemented with non-coherent optical sources and, with sufficiently high modulation, displacement changes of some tens of micrometres are possible. Additionally, there is no limitation for the maximum value of d, except what derives from collimation and optical loss issues.

5.4. Resolution Limits in Optical Displacement Sensing

Optical measurement is normally associated with high performance, being understood by that the ability to reach extreme resolutions when applying optical techniques in the measurement of a particular parameter. Displacement is one of the parameters where the perception of such performance is more accessible, in view of our daily experience with movement. Therefore, we consider useful to include in this chapter some information about what can be achieved nowadays with the available optical technology for displacement measurement and provide some thoughts about future developments.

5.4.1. Resolution Limit in Classical Sensing Systems

In classical optical sensing systems the ultimate resolution comes from the realization that the light intensity cannot be measured with infinite precision in view of its fluctuation around some average value. To quantify this effect, we can start from the Heisenberg uncertainty principle applied to the pair energy-time, namely $\Delta E \cdot \Delta t \geq h/2\pi$, where ΔE and Δt are, respectively, the uncertainty in the measured energy and the uncertainty in the time window within which the measurement is done, with h the Planck constant. If n is the average number of photons in the light field (assumed monochromatic), then $E = n\omega h/2\pi$ and $\Delta E = \Delta n\omega h/2\pi$, where ω is the light angular frequency and Δn is the fluctuation in the detected photon number. At a specific point in space, the light phase is given by $\phi = \omega t + cons\tan t \Rightarrow \Delta\phi = \omega\,\Delta t$ with $\Delta\phi$ the fluctuation of detected optical phase. The insertion of these relations into the energy-time uncertainty condition gives

$$\Delta n\,\Delta\phi \geq 1 \qquad\qquad (5.33)$$

known as the Heisenberg number-phase uncertainty relation.

A single frequency laser light field is well approximated by a coherent state, commonly represented by $|\alpha\rangle$, with $\alpha = |\alpha|e^{i\phi}$ is proportional to the electric field amplitude A. Therefore, $|\alpha|^2$ is proportional to n and thus to the intensity of the light field. The intensity can be expressed as $I = nI_{sp}$, where I_{sp} is the intensity associated to a single photon, given by $I_{sp} = h\omega/(2\pi\varepsilon_0\eta)$, where ε_0 is the vacuum electric permittivity and η is the mode volume of the electromagnetic field. The statistics of the light intensity fluctuations follows a Poisson distribution, meaning that $\Delta n = \sqrt{n}$; also, for a coherent state light field, $\Delta\phi = 1/\sqrt{n}$, which means equality is considered in equation (5.33), indicating that such light state is a minimum uncertainty state, with the relevant relations summarized as:

$$\Delta n\,\Delta\phi_{sn} = 1$$
$$\Delta n = \sqrt{n} \tag{5.34}$$
$$\Delta\phi_{sn} = \frac{1}{\sqrt{n}}$$

The third equation of this group can be expressed in terms of intensities in the form

$$\Delta\phi_{sn} = \left(\frac{I_{sp}}{I}\right)^{1/2} \tag{5.35}$$

known as the shot-noise limit for the determination of the field light phase. This phase uncertainty can be expressed as an equivalent length uncertainly, $\Delta\ell_{sn}$, through the spatial phase relation $\Delta\phi_{sn} = k\,\Delta\ell_{sn}$, where the wave vector k is $k = 2\pi/\lambda$, with λ the light wavelength in the medium. From the third equation (5.34) it results

$$\Delta\ell_{sn} = \frac{\lambda}{2\pi\sqrt{n}} \tag{5.36}$$

the *minimum displacement change* that can be measured in a shot-noise limited interferometric sensing system, therefore establishing the resolution limit of classical sensing systems when addressing displacement measurement.

It is instructive to evaluate $\Delta\ell_{sn}$ for the situation of one of the most ambitious scientific projects going on nowadays, the *Laser Interferometer Gravitational-Wave Observatory* (LIGO), a scientific collaboration of the California Institute of Technology and the

84

Massachusetts Institute of Technology, established in 1992, with infrastructures located at Hanford, Washington, and Livingston, Louisiana. It aims at the detection of gravitational waves, ripples in space-time induced by violent astronomic events, like the explosion of a supernova. When these waves propagate they induce dimensional changes in the distance between two objects, which can be sensitively detected with interferometric techniques. The LIGO apparatus consists of two Michelson interferometers with four kilometre long arms. A schematic of the LIGO interferometer is shown in Figure 5.12 (the two places in United States where the two interferometers are installed are also shown).

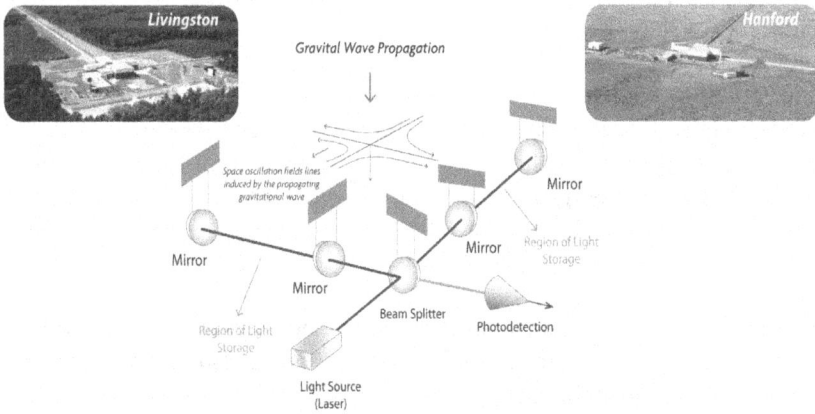

Figure 5.12. Layout of the LIGO Michelson interferometer (inset: photos of the two places where interferometers are installed).

The optical power circulating in these interferometers is in the order of 100 kilowatts with wavelength of $\lambda \approx 1\,\mu m$, which corresponds to a mean photon number $n \approx 10^{24}$. Therefore, it is expectable a shot-noise determined minimum detectable displacement of the interferometer mirrors of $\Delta \ell_{sn} \approx 1.6 \times 10^{-17}\,m$.

This value is roughly a thousand times smaller than the diameter of a proton! Surely, it is quite impressive but nowadays we know it is not the ultimate limit. The consideration of particular light states, the so-called squeezed states, offer the possibility to go well further, as briefly outlined in the next section.

5.4.2. Squeezed Light and Displacement Sensing

The shot-noise phase uncertainty expressed by equation (5.35) results from the photon-number fluctuations associated to a light field. In a classical electromagnetic field perspective, the field amplitude and phase can be measured simultaneously with infinite precision, meaning with proper design and experimental effort it would be possible to reach smaller and smaller minimum detectable phases (and detectable lengths) in interferometric sensing systems (the same would happen with sensing based on different modulation approaches, such as those supported by intensity light modulation – intensity sensors – where now the relevant phenomenon would be the intensity fluctuations). The quantification of the light field implied such picture was no longer valid and, for some time, it was assumed that the shot-noise effect determined the ultimate resolution as imposed by quantum mechanics. The situation changed in 1981 with a breakthrough paper of Carlton Caves, where it was proposed the idea of using non-classical states of light to improve the resolution of optical measurement systems to even below the shot-noise limit. These non-classical states were identified as squeezed states of light and the following paragraphs provide a glimpse of their main characteristics.

A light field in the classical perspective has no fluctuations; therefore it is represented by a point in a particular representation known as *phase space*. A coherent light state (like a laser light field) is shown by a disk indicating that we are in presence of intensity and phase fluctuations of the same amplitude. If the area of this disk is A, equations (5.33) and (5.34) indicate any light quantum state must have an area greater or equal to A; also, any minimum uncertainty light state must be represented in the phase space by a region of area A. In other words, while for the coherent state the three relations of equation (5.34) are fulfilled, minimum uncertainly light states are possible by relaxing the last two relations and keeping valid the first one.

The consequence of this argument is that it is possible to decrease $\Delta\phi$ when it is permitted to increase Δn, so that the product $\Delta n \Delta \phi$ remains unitary and the area of the enclosed uncertainty region keeps equal to A, indicating the fulfilment of the condition for having a minimum uncertainty light state. In the phase space this is represented by squeezing the disk associated to the coherent state into an ellipse such that the phase uncertainty is decreased at the expense of increasing the light intensity uncertainty.

A simple argument permits to estimate how much squeeze can be performed on a light state. It is based on the question of what would be the largest uncertainty that can be produced in the intensity of a light state with a mean photon number n while keeping the minimum uncertainty state condition, i.e., the area of the ellipse remains equal to A. It is fairly obvious that the energy fluctuations of the light state cannot exceed its average energy, which is equivalent to say that at maximum $\Delta n = n$. When this relation is inserted into the first equation of the set (5.34) we get for the ultimate phase uncertainty of the light field

$$\Delta \phi_{\mathrm{HL}} = \frac{1}{n} = \frac{I_{sp}}{I} \tag{5.37}$$

This relation expresses what is known as the Heisenberg Limit, designation justified by the fact equation (5.37) gives the minimum phase uncertainty possible for light with an average photon number n.

In an interferometer this phase uncertainty is converted into a displacement uncertainty given by $\Delta \ell_{\mathrm{HL}} = \lambda / (2\pi n)$. Following the example of LIGO indicated before, for an average number of photons $n = 10^{24}$ circulating the interferometer and $\lambda = 1$ µm it comes out , not far away the Planck length limit (10^{-35} m) where the classical notion of space breaks down. This improvement of twelve orders of magnitude in the interferometric phase resolution relative to the one determined by shot-noise is not expectable to be reached in the laboratory in the foreseeable future due to technological limitations.

Squeezed light can be generated from light in a coherent state by using certain optical nonlinear interactions, such as optical parametric amplification and frequency doubling. The Kerr nonlinearity in optical fibres also allows the generation of squeezed light, as well as semiconductor lasers when operated with a stabilized pump current. Gradually, the achievable noise reduction through light squeezing is increasing, a value of 10 dB being reported in 2008 and of 12.3 dB at 1550 nm in 2011. The *Advanced Ligo Project,* an upgrade of LIGO nowadays being implemented, relies strongly on the application of squeezed light in the interferometer to substantially decrease the level of interferometric noise.

The quantum states of light associated with coherent and squeezed states deal with a large number of photons and do not consider the deep phenomenology intrinsic to quantum mechanics, the entanglement phenomena that couple in a non-local way different elements of a quantum system once they are allowed to interact. In some way, such system keeps an interferometric pattern instantaneously and strongly sensitive to a perturbation in any part of the system, independently of how far it is from the other system elements. The definition of a quantum sensor as a quantum mechanics system projected to probe the quantum state of the system under measurement relies on such entanglement mechanism.

It is fairly obvious to realize we are facing a radically new optical sensing approach that promises unthinkable performance. It is an historical fact that optical sensing, particularly fibre based, benefits substantially from the optical communications endeavour, a huge enterprise that brought the physical infrastructure of today's information society, providing concepts and technology that permitted a fast development of optical sensors. It is conceivable the same will happen with quantum optical sensing, now with the locomotive being the whole field of all-optical quantum computing. Indeed, an optical quantum computer can be thought of as a giant optical quantum interferometer with many arms and a large but discrete number of photons circulating through it, with the quantum entanglement between photons programmed, for example, to carry out mathematics impossible in any classical computer machine; the sensing approach is supported in the same concept, not to solve complex mathematical problems but, instead, to perform ultrasensitive detection of a specific measurand.

The concept associated to the operation of a quantum optical sensor indicates that the privileged configurations will be of interferometric nature. Their research and development involves complex issues such as the design and set up of optical sources of entangled photons for two-mode (two paths) or multi-mode (multipath) layouts, the conception of effective mechanisms of measurand interaction with the entangled photon set, the conversion in the interferometric output of the measurand modulated photon entangled state into an optical intensity signal in a form compatible with the application of standard optical interrogation techniques, the identification of factors that degrade the photon entangled condition with the consequent degradation of sensor performance, and many others.

When this concept of quantum optical sensing is applied to the measurement of displacement, the point that needs to be emphasized here is that the possibility of unthinkable resolution levels by the current standards will become theoretically feasible, opening paths for their experimental validation first, then for their application as a probe to explore new frontiers in science, to open new technological territories and, at a later phase, to be incorporated into engineering, all of this indicating that fascinating times wait us ahead.

5.5. Concluding Remarks

Displacement is one of the core entities of the physical world. Objects move relative to each other, their interactions and associated forces are dependent on their distance; in a system the relative displacement of its constituent elements generate reactive actions, and so on. This means that both in science and technology the ability to measure displacement is crucial, which justifies the efforts done along the ages to perform displacement measurement with the best possibly accuracy and reproducibility. Many approaches are possible to carry out displacement measurement, but for long optical techniques have been considered as synonym of high performance and, therefore, they have been considered with particular focus. With the advent of the optical fibre, additional possibilities emerged within the framework of optical fibre sensing and this chapter aimed to present some of the associated developments.

We started by introducing a broad classification of fibre optic displacement sensors with the *contact* characteristic as the identifying element. Two types of contact displacement sensors were introduced, the microbend displacement sensing structure and the spring interfaced optical displacement layout, the latter being the object of a general theoretical analysis to identify its main characteristics. Optical techniques are often associated with contactless operation; therefore, several optical fibre sensing configurations for non-contact displacement measurement were presented. Some of them were analyzed in detail to assess their main characteristics.

Within the context of optical displacement sensing it was considered relevant to address the resolution limits issue. We started by dealing with this topic in the realm of classical physics and later moved on to discuss what becomes possible when semi-classical light states, such as squeezed light states, are introduced. To illustrate the impact of optical techniques for displacement measurement when cutting edge performance is demanded, a brief description was made of their application in the LIGO project that aims to detect gravitational waves with advanced optical interferometry. Finally, some thoughts were delivered on the fascinating new territories of quantum sensing that will lead to developments not imaginable nowadays.

Bibliography

Abadie, J., B. P. Abbott, R. Abbott et al. 2011. *A Gravitational Wave Observatory Operating Beyond the Quantum Shot-Noise Limit: Squeezed Light in Action.* arXiv:1109.2295v1 [quant-ph].

Berthold, J. W. 1999. *Microbend Fiber Optic Sensors.* Springer.

Bruce J. Torby, *Advanced Dynamics for Engineers*, 426p, Holt-Saunders Int. Ed., 1984.

Cavaleiro, P. M., A. B. Lobo Ribeiro, J. L. Santos. 1995. *Referencing Technique for Intensity-Based Sensors Using Fibre Optic Bragg Gratings.* Electronics Letters 31: 392-394.

Caves, C. M. 1981. *Quantum-Mechanical Noise in an Interferometer.* Physical Review D 23: 1693-1708.

D. S. Nyce, *The LVDT*, in Linear Position Sensors, Vol. 98, pp. 94–97, Wiley-Interscience, Hoboken, NJ, USA, 2004.

Dawson, A. *Inductive Distance and Movement Gauge.* UK Patent No. 9116251.1, 1992.

Dowling, J. P. 2008. *Quantum Optical Metrology – The Lowdown on high-N00N states.* Contemporary Physics 49: 125-143.

Ernest O. Doebelin, *Measurement Systems, Applications and Design*, MacGraw-Hill, 1990, ISBN 0-07-100697-4.

Frazão, O., L. A. Ferreira, F. M. Araújo, J. L. Santos. 2005. *Applications of Fiber Optic Grating Technology to Multi-Parameter Measurement.* Fiber and Integrated Optics 24: 227-244.

Georges Asch, *Les Capteurs en Instrumentation Industrielle*, Dunod, 1991, ISBN 2-10-000220-1.

Gloge, D. 1972. *Optical Power Flow in Multimode Fibers*. Bell Systems Technological Journal 51: 50-59.

Grattan, K. T. V. and B. T. Meggitt. 1995. *Optical Fiber Sensor Technology*. Chapman & Hall.

James W. Dally, William F. Riley e Kenneth G. McConnell, *Instrumentation for Engineering Measurements*, John Wiley and Sons Inc., 1993, ISBN 0-471-55192-9.

John G. Webster, *The Measurement, Instrumentation, and Sensors Handbook*, CRC Press LLC, 1999, ISBN 0-8493-8347-1.

Kapale, K. T., L. D. Didomenico, H. Lee et al. 2005. *Quantum Interferometric Sensors*. Concepts of Physics 2: 225-240.

Kolkiran, A. 2009. *Quantum Imaging and Sensing with Entangled Photons: Heisenberg Limited Measurements with Entangled Light.* VDM Verlag.

Konwar, H. and J. Saikia. 2012. *On the Nature of Vacuum Fluctuations and Squeezed State of Light*. Archives of Physics Research 3: 232-238.

Ladd, T. D., F. Jelezko, R. Laflamme et al. 2010. *Quantum Computers*. Nature 464: 45-53.

Lee, B. H., Y. H. Kim, K. S. Park, J. B. Eom, M. J. Kim, B. S. Rho and H. Y. Choi. 2012. *Interferometric Fiber Optic Sensors.* Sensors 2012: 2467-2486.

Linear Displacement Measurement with Eddy-Current Sensors, LION Precision Application Note LA02-0060, March, 2013.

López-Higuera, J. M. (Editor). 2002. *Handbook of Optical Fiber Sensing Technology*. John Wiley.

Lorenzo Sciavicco e Bruno Siciliano, *Modeling and Control of Robotic Manipulators*, 358p, McGraw-Hill Int. Ed., 1996.

M. Repetto and J. Simkin, *Engineering analysis for design optimization of differential transformers*, Computer-Aided Engineering Journal, Vol. 5, No. 2, pp. 51–53, 1988.

Mehmet, M., S. Ast, T. Eberle et al. 2011. *Squeezed Light at 1550 nm with a Quantum Noise Reduction of 12.3 dB.* Optics Express 19: 25763-25772.

Potapov, V. T., D. A. Sedykh and A. A. Sokolovskii. 1988. *Fiber-Optic Interferometric Displacement Sensor.* Measurement Techniques 31: 561-563.

R. S. Weissbach, D. R. Loker, and R. M. Ford, Test and comparison of LVDT signal conditioner performance, in *Proceedings of the IEEE Instrumentation and Measurement Technology (IMTC '00)*, pp. 1143–1146, May 2000.

Santos, J. L, and A. P. Leite. 1990. *Estudo do Sensor de Reflexão em Fibra Óptica.* Departamento de Física da FCUP.

Scully, M. O. and M. S. Zubairy. 1997. *Quantum Optics.* Cambridge University Press.

Vahlbruch, H., M. Mehmet, S. Chelkowski et al. 2008. *Observation of Squeezed Light with 10-dB Quantum Noise Reduction.* Physics Review Letters 100: 033602.

Vincent Janoo, Lynne Irwin, Kurt Knuth, Andrew Dawson and Robert Eaton, *Use of inductive coils to measure dynamic and permanent pavement strains*, accessed May 2013, http://www.mrr.dot.state.mn.us/research/apt/data/cs03-02.pdf

Yasin, M., S. W. Harun and H. Arof (Editors). 2012. *Fiber Optic Sensors.* InTech.

Zheng, J. 2004. *Analysis of Optical Frequency-Modulated Continuous-Wave Interference.* Applied Optics 43: 4189-4198.

Index

A

absolute encoders, 53, 54, 57, 61

accuracy, 56, 57, 61

Acoustic pulses, 21

active coil, 48, 49, 50, 52

amplitude modulation, 40, 42

angular displacement, 12, 53

beam, 21, 59, 68, 69

beat frequency, 69, 84, 85

B

BIPM, 16, 17

Bragg wavelength, 72, 83

C

calibration, 52, 70

capacitance, 7, 19

Capacitive displacement transducers, 19

characteristic, 45, 47, 93

collimation, 68, 70

conducting film, 24, 25, 30

contact fibre optic displacement sensor, 64, 71

coupling efficiency, 78, 80, 81

D

demodulator, 44, 45

digital encoders, 9, 10, 22

displacement, 11, 26

displacement range, 66, 71

distance, 9, 11-13, 16, 17, 19, 20, 48-52, 59, 64, 67-71, 73, 74, 80, 83-86, 88, 92

dynamic response, 47

E

eddy current, 20, 48, 52

encoders, 53

F

FBG, 71, 72

ferromagnetic core, 19, 34, 35, 37, 40, 45, 47

Fibre Bragg Grating, 64

fibre optic displacement sensors, 63

fibre optic intensity sensors, 68

Foucault current, 48

Frequency Modulation Continuous Wave, 84

frequency response, 52

Frictionless, 46

fringe, 68

G

Gray code, 56

H

hybrid potentiometric transducers, 25

I

incremental encoder, 58

incremental encoders, 53

inductance, 7, 19, 20, 36, 49, 50

inductive displacement transducers, 19

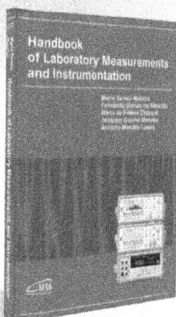

www.ingramcontent.com/pod-product-compliance
Lightning Source LLC
Chambersburg PA
CBHW060931240326
41458CB00139B/1414